Surrogacy and Other Reproductive Technologies

A. Ramija Bee
N. Manimekalai

Surrogacy and Other Reproductive Technologies

A. Ramija Bee
N. Manimekalai

Academica Press
Washington~London

Library of Congress Cataloging-in-Publication Data

Names: Bee, A. Ramija (author)
Title: Surrogacy and other reproductive technologies | A. Ramija Bee
Description: Washington : Academica Press, 2022. | Includes references.
Identifiers: LCCN 2022942141 | ISBN 9781680538786 (hardcover) | 9781680538793 (e-book)

Contents

Chapter I

Introduction

Background

The human civilization has tremendously changed, with the advent of science and technology. Technologies *per se* do not harm any living being but its application does affect the human living. Human beings live in a complex society, rules, where the constructed by socio cultural norms are adhered to by various social groups.

It has never been questioned rationally and scientifically, as to why, who, for whom such socio-cultural norms were constructed and implemented, which restrain equality of human beings. The institution of family has been the first breeding ground for the discriminative and biased social norms, which, coupled with other institutions like caste, class, religion, race, language, educational institutions, literature, media, market, State, etc, reinforce the gendered social norms to be followed by various social groups. However, history has evidenced that the family structure is being dramatically changing with alternative perspective and even questioning or countering the existing norms. Human being is in search of knowledge, to help to evolve into higher levels of evolution.

Survival of the human species is the most important of all the developments and hence the need for procreation. Procreation is defined as producing progeny. Each and every organism is producing its off-spring. Reproduction is a biological process and there are two kinds of reproduction sexual and asexual. Sexual reproduction is mainly carried out by the humans and mammals. Human reproduction is mainly gynocentric. Due to this gynocentric contribution, women are often recognized and identified with their reproductive roles, though women can play other roles. The scientific focus of viewing reproduction as just a part of the women's lives is missing. It is the need of the hour that women should not be just associated with the reproductive role alone because several women make their choice to stay single or marry and choose not to have a child or adopt a child. But the social expectation that marriage and child birth is the ultimate for women, often reduces woman to be a mere mother.

Women failing to reproduce were ill treated in the family and social system and they were put into extreme humiliation and infertile woman were socially excluded. Infertile women were not allowed to participate in ceremonies and stigmatized for the failure to procreate and have a heir to the family in the Indian social system. Manu permitted men to remarry, after ten years of living with an infertile wife. Similarly, the patriarchal expectation that if she does not give birth to a male child, men can go for second marriage for the sake of having a son that only a male heir can. People still believe carry forward the lineage of the family. In the social system, prevailing the system of patrilocal marriages are the norm and women are made to move from maternal home. Home of orientation with such arrangement expecting only son to be the carrier of family is quite unfair and renders injustice to women.

The religious expectation further shaped and guided the caste system in India to develop such expectations, where every caste has its own norms to handle infertility. A certain commercial class in Tamil Nadu has organized adoption from their own community when someone does not have biological child. Such arrangements are hardly seen in the mainstream system.

The very expectation that women are born to reproduce and failing which she is not fit to lead a peaceful life, indicates the subjugation and subordinate status of Indian women. Family and caste patriarchies shape the norms and often keep the women in secondary status, with the series of social sanctions for men and restrictions on women. Gender justice cannot be realized with such a divided society. Adoption laws are universal in the West but in India, it is applicable only to Hindu religion, as per the Hindu Adoption and Maintenance Act of 1956. Increasing infertility due to changing life style, declining children for adoption, procedures to be followed in adoption etc, prompted barren women to seek other ways to get a child.

Infertility in India

Infertility is scientifically defined as at least 12 months of unprotected intercourse, without conception. There are many reasons for infertility such as early marriage, nutrition, diseases, life style changes and malformation of the uterus. Female infertility is at 40-50 percent and male fertility is at 30-40 percent. Nearly, 27.5 million couples, in India, are affected by infertility (TOI, 2017).

It was generally believed in the society that a woman, who is not able to procreate, is not fit for life and hence she could be excluded from the family structure. In the process, these social and cultural taboos kill and humiliate the woman emotionally and psychologically. In the daily life, infertile women face different kinds of struggles and sufferings. It cannot be measured how much stress, pressure and humiliation women face during every menstrual cycle, to expect to conceive but failed. There are reports, by the Doctors who handle infertility that family members are least bothered about women's health and they are prepared to sell off the properties, to have a child, a legitimate heir. It is further added that with such struggle, some couples may succeed in getting a child. In case the baby were to be female, it would further be humiliating to women as if the sex of the child were to be determined by the woman. Dr. Charmila a leading Gynecologist in Tiruchirappalli, Tamil Nadu reported that there was no happiness in the birth of a female child. It was further added that the birth of a baby girl caused delay to the payment to be made to the Hospital and a gloomy and sad environment prevailed. All these issues, from a feministic and humanistic perspective, must be interrogated scientifically to find out whether it is the fault of women to have a baby girl. Customary laws and social norms define the negative value of a girl child. The society does not entertain and enjoy the birth of a baby girl for the reason that it will not ensure the future insurance for family. Even the male children do not support the parents in future, but still there is socio cultural expectation that the funeral rites are to be performed by male heir. All these really pressurize the infertile couples to opt for a male child, reinforcing the stereotyped values over male preference. It is irrational, unjust and unethical also to treat the female as subordinate and continue the suppression and oppression of girl children despite the development in science and technology, education and empowerment.

The New Assisted Reproductive Technologies have been serving the infertile couples to have their own genetic child. The ARTs are emerging as the panacea, for infertile couples, across the countries in the world. As literature revealed that it serves to help the homosexual partners also to have a child. The earlier adoption scheme had stringent procedures and did not allow the couples to have their own genetic child. Hence parents preferred their own biological child and consequently, there has been a huge market, emerging for such ARTs, including surrogacy.

On the supply side, the economically backward and vulnerable sections are being targeted by the intermediaries. The price for this vulnerability is that women are exploited in terms of poor payment and they are abandoned once the child is delivered. It is inhuman to just leave the mother, who carried the child in her womb and hence at least one-year of health insurance may be made part of the payment. As demand grows and with the stringent procedure towards the adoption of a child in various other countries, coupled with huge cost involved, India has figured as a potential market. Women, who opt for surrogacy, are often not informed about the dynamics involved in the market, especially the black market because the parents, who look for a child, do not directly come in contact with the surrogate mothers. Moreover, there is a requirement to maintain confidentiality and not to reveal the identity of the surrogate mother, to the couple, seeking a child. Against this background, the question of undertaking research on social, economic and political issues, related to ARTs in general and surrogacy in particular, posed great problems and as such, research on such issues has been very limited. Several researches have been undertaken, focusing on scientific and technological perspective but not from humanistic and gender perspective.

In the contemporary context, the face of families has been changing dynamically, with the development in economic terms, where the economic reforms in the form of liberalization, privatization, capitalism, and globalization, have taken the people to the height of consumerism. The traditional masculine and patriarchal notions of family and kinship enforced expectations on women, to procreate biologically. The procreative capacity created inequalities, subordination, and discrimination in the society. Now with the advancement in reproductive technologies, there has been a trend to move from marital fertility to non-marital fertility, with the help of reproductive and Assisted Reproductive Technologies. In this context, the children, born outside marriage, may be born to cohabiting couples or to a single woman, either through IVF or other methods with the application of Assisted Reproductive Technologies.

Assisted Reproductive Technology Market - Feminist Stand point

As already mentioned, infertility is growing and it is estimated to affect 48.5 million duos globally (WHO, 2010). It is generally agreed to be a 'major life crisis.' Emotionally devastated, many infertile duos follow fertility drugs, surgery and new reproductive technologies, in their hopeless attempts to have children. Gynecologists have recently developed several new methods, for the remedy of infertility. *In Vitro* Fertilization (IVF) and other high procreative technologies, where conception mainly takes place in a laboratory, have become the present focus of medical research and they are now consistently recommended to infertile duos by gynecologists. Expansions of the medical profession, into new areas of women's health, have been treated with suspicion by feminist scholars in recent years (Ruzek 1979, Gordon 1977, Ehrenreich and English 1978), and many feminists have been alarmed by the new reproductive technologies. To Gena Corea, member of the Feminist International Network Resisting Reproductive and Genetic Engineering (FINRRAGE), 'the issue is not fertility but it is the exploitation of women.' Many feminist biologists, sociologists and philosophers consider IVF, embryo freezing, egg donation, surrogacy and other new developments, in infertility treatment, harmful and threatening to women's right to control their procreation. In fact, the degree of consensus among feminists on the issue of reproductive technologies is occasionally found on other areas also. Internal dispute, on the fundamental theoretical issue of the 'centrality of motherhood' in women's lives, has not produced conflict over the acceptability of IVF but instead has led to discrepancy over the primary reasons for criticizing these technologies. In a way, the ARTs reinforce the motherhood which may exclude the preferences and choice of single women, and gender non-conformity groups such as LGBQT etc. The choice and concerns of such women cannot be excluded by glorifying motherhood. In the extreme case, it compelled the women to seek the help of technologies, which often do not turn successful but the pain that women face during the treatment process, cannot be expressed adequately.

Baby Selling, Victimization and Surrogacy

Defenders of commercial surrogacy also carefully distinguish it from baby selling. Children are not sold as commodities, but rather women's reproductive services are for sale. Since men have been allowed to sell their sperm, why should women be prevented from participating in an analogous transaction? Finally, defenders point out that commercial surrogacy offers new ways for gays and lesbians and single people to become parents.

Feminist Gena Corea (1985) argued that the reproductive technology is the product of male reality, which had also become the social reality. The values, expressed in the technology like objectification and domination, are typical of the male culture. The technology is male generated and buttresses male power over women.

Women's Freedom of Choice and Surrogacy

Some feminists see women's subordination, to be fundamentally caused by their role in reproduction and according to them, only test tube babies will make gender equality possible (Firestone 1970). Feminists are divided over the issue of commercial surrogacy. Those, who support surrogate motherhood, often stress the increase in the freedom it brings. Surrogate contracts allow women to have additional choices over their reproduction. Carmel Shalev (1989) goes further and argues that prohibiting such contracts, fails to give due respect to the choices women could make. If a woman freely enters into a contract to produce a child, it is paternalistic and demeaning to prevent her choice. Last few decades, India has seen radical revolution in the fertility services. In the trendy era, India is seen as a paradise for infertility treatment. Because of globalization, the number of infertility cases has increased every year. Assisted Reproductive Technologies plays a crucial role and India has become very famous in the "Fertility Industry." Because of this fertility industry in India, Medical Tourism has become fashionable at the global level. In India, due to the success rate of this industry, the medical tourism has morphed into reproductive tourism. A.U.N. - backed study, in July 2012, estimated the surrogacy business at more than 400 million USD a year with over 3,000 fertility clinics across India. Majority of the countries forbid commercial surrogacy due to economic exploitation and some countries prohibit all forms of surrogacy.

Technology, Surrogacy and Motherhood - Feminist Critique

Thanks to technological development and advancement, particularly reproductive

reproductive technologies, which have served the needs of the couples to have children, the Assisted Reproductive Technologies, could be a blessing to the infertile couples. Surrogacy enables the infertile couples, to get a biological or genetic baby, but the question is at what cost.

Historically speaking, the reproduction was not an issue as long as people lived in harmony with nature and there were no external interventions, to prevent the natural process of maturity to reproduce, be it women or men. Infertility was not much prevalent, though there was incidence of polygamy, where men had the social sanction for adultery and had sexual relationship with more than one woman. Patriarchy based, male written norms had been accepted as social norms, without any dispute. The Hindu Religious Law, the Manu Smirti, openly permitted formal marital relationship for the sake of getting a male child. Women or any sensitized man never interrogated such gendered social norms, which are quite biased and continue till date in the form of female feticide, with the help of technology, where the second girl baby was not much welcomed, thanks to the son preference being the norm till today.

In the human race, reproduction plays a major role because procreation alone gives meaning to human life. The human family structure and culture depend on the offspring. Incidentally, masculinity and patriarchy are strongly correlated with procreation. The concept of masculinity and patriarchy forced women to produce children. Men and women, who were not able to procreate, were excluded and ill-treated in the society. Infertility destroyed the entire family structure of human beings.

Feminists critiqued this patriarchal value because it reduced pregnancy to mere service and the baby to a product. There is a market for surrogacy women, from poverty ridden households often fall prey to. Originally, it was taking place with family, relatives, friends and it was not commercial. With the scientific advancement and technological development, even couples from foreign countries travel to India, to hire a surrogate mother because in their countries it is banned, or involves huge legal formalities and cost. The readiness of the surrogate market, at the cost of the exploitation of poverty and due to the economic backwardness of women, relatively low cost in India in producing a baby, the emerging market with competition, less or no regulation etc, facilitated the foreign couples to get the baby through assisted reproductive technologies, particularly surrogacy.

Statement of the Problem

The question of infertility and its associated patriarchal expectation that it is the women, who should handle the issue, all push women to this state of vulnerability. Women do not get the choice to remain childless or go for adoption. Rather her body is being completely put into scrutiny, with medical equipments, where the humanity is lost at the altar of expectation in the society. Is woman not a human? Is she a machine? Can’t a woman be allowed to remain childless, though it is not her choice but the nature's fault or the defect of the husband? Thanks to the reproductive technological advancement, the woman could choose the technology to get a baby and that too, a biological baby. Initially, couples were enough to procreate a child but in the current era, one needs the third person to procreate. The third party needs help from technologies to procreate and hence the procreation is termed facetiously as "technologies of procreation." At this juncture, the infertility or sub fertility people were motivated to use technology, to have a biologically related child. There are a number of technologies but compared to other techniques, surrogacy is mainly preferred by the duos, to get a biological child. Meanwhile, surrogacy creates a lot of problems in the society, especially to women. In addition, lack of research on surrogacy also poses a problem, for government agencies, to take substantive legal action against those found guilty and a number of surrogacy related questions remain unanswered. Instead of adoption, surrogacy is currently on the public agenda because of increasing desire to get their biological child. Before technological evolution, adoption has always been a potential means, for an infertile couple, to become parents. Evolution of Reproductive Technologies, helped couples to overcome social stigma of the infertility and to fulfill their desire obtain respect from the society in the process. The Assisted Reproductive Technology, as an opportunity for family formation, recognizes that the commercial and social components of mothering can be separated and accomplished by different persons.

Theoretical Background

Infertility has been a social issue for long, where couples, who could not get a child, preferred adoption or medical treatment. Historically speaking, there were no technologies to help infertile couples to get a child. In mythology, there were several instances of couples getting their biological child, by having their sperms donated to other women and getting the baby conceived. Surrogacy was in practice in the ancient period as mentioned in the epics. With the advent of science and technology and medical innovations, there are different Assisted Reproductive Technologies in which surrogacy is one form, which provides a biological child, to the commissioning parents.

There are two types of surrogacy; traditional and gestational. Traditional surrogacy is done through artificial insemination, where the surrogate uses her own egg with another man's sperm. Gestational surrogacy is done by *In-Vitro* fertilization, where the fertilized eggs are implanted in the uterus of the surrogate. There are preferences for the later method because it gives a full biological child compared to the traditional method where the parent can claim only half sibling. Hence there is a natural preference for the IVF methods.

Infertility is treated with the advent of the technology but still there are issues as to why infertility cannot be accepted by the society. With such a perspective, should the technology in terms of ARTs, be entertained which, though they help the childless couple to have a child, they also reduce women's body to a reproductive machine? Can the society not sustain without procreation of a miniscule percentage, who are not able to procreate? Will it not be the responsibility of men and members of the family to advocate for adoption, which would still be rewarding? Can the women not be spared from the social pressure to reproduce? Do members not be concerned with physical and psychological pains of women with infertility, while undergoing the treatment for infertility and these are the questions to be resolved, while doing research on surrogacy.

Research Questions

The present study raises certain basic questions like under what circumstances the couple for the fertility treatment in general and surrogacy in particular, the term and conditions laid down, the legalities in hiring a surrogate mother, how far the rights of the surrogate mothers are being ensured, if not what are the basic concerns and challenges that the surrogate mothers face, what is the background from which surrogate mothers are being drawn, how comfortable the surrogate mothers are in convincing the family, withstanding the pressures from commissioning couples, facing the post delivery issues, are the agreed terms and conditions, in terms of payment to be made to the surrogate mothers, fulfilled and. If not, what are the issues, how best such issues be tackled legally, is there any exploitation of the economical backwardness of the surrogate mothers, by the stakeholders involved in the surrogate market arrangements particularly, the agents and intermediaries, what are the general health issues that the surrogate mothers face after handing over the child, what are the legal safeguards? How does the recent Surrogacy Bill/Act help or affect surrogacy? Is the new bill accommodate all concerns of surrogacy in a larger perspective? All these questions are being analyzed in the present study with the support of primary and secondary data, and also oral interviews, with all stakeholders concerned. It is to be underlined that the present study would focus on the implications of Assisted Reproductive Technologies (ARTs), especially surrogacy on women and their reproductive role from a feminist perspective, the factors and sources which have driven women to serve as surrogate mothers, the intrusion of the surrogate market, taking advantage of the economic vulnerability of the women. Given that the surrogate market offers a considerable economic relief to households assuring a lump sum, it raises a fundamental question as to whether the woman's body or uterus be targeted, reducing women to only the reproduction and whether such role be commercialized for the sake of household economic needs but glorifying that as a sacrifice, service to commissioning parents, unfolding that pains and pressures that the surrogate mothers face pre and post surrogacy arrangements. Capitalist patriarch making women themselves, to speak out as a service and not a pain or pressure, is the social reality.

Objectives

Based on the above research questions the following objectives were framed.

- ✓ To study the socio economic, demographic characteristics of the Surrogate Mothers.
- ✓ To identify the motivation, obstacles and facilitative sources to offer to surrogacy.
- ✓ To analyze the process of identification, persuasion and compromise the women make, the nature of market through which intermediaries for surrogacy emerge, and to locate the associated issues, the gap, if any, between the terms and conditions agreed upon and actually implemented in a legal perspective.
- ✓ To find out the issues faced by the surrogate mothers post surrogacy – emotional, psychological, physical health and trauma and strategies to manage the same.
- ✓ To present the perception of various stakeholders like Doctors, Associations, Surrogate Mothers, Community and present the reflections on the New Draft ART Bill 2016, with the pre and post bill changes and safeguarding mechanisms.
- ✓ To provide suggestions and recommendation based on the findings, towards policy formulation.

Significance of the Study

India has the largest surrogate market in the world. Surrogate mothers in India are drawn from diverse background and majority of them are economically and socially backward drawn from families. Majority of the surrogate mothers opted for surrogacy only to pay off their past debts in their households. Despite the existence of conventions and legislation on the subject, surrogate mothers increased substantially in recent years. However, the issue of surrogate mother in India is receiving a lot of attention, both from the Government and Non - Governmental Organizations, but the extent and context of surrogate mothers in India, is too vast to be tackled by any single organization. There are number of questions to be discussed. Is surrogacy legally bound? What will happen if the surrogate mother refuses to continue after the agreement? Will the payment be made by the commissioning parents as agreed? What would be the consequences if the conception fails repeatedly? If commissioning couples look for surrogate mothers, what is the process, how do they identify the market, how do they get their immigration clearance? What would happen in case the legislations are passed to regulate and restrict surrogacy? Will it have implications for illegal market? What is the height of the problem of surrogacy? What priority to be accorded to the New Legislations to regulate surrogacy? Often the rights of the surrogate mothers, child and the commissioning couples have been put into troubles and any regulation need to accommodate these concerns.

The present study would enhance the development of relevant literature, for realistic planning and understanding surrogacy and surrogate mothers in India, especially in Tamil Nadu. This study would also help in contributing constructive solutions, to address their problems and needs. From the feminist point, surrogacy has implications on the status of women. Women have been respected for the 'role of reproduction' in majority of contexts. If uterus could be hired out, it might deteriorate the status of women. From the literature, it is understood that women from economically backward class, opt for surrogacy. There is concern that if reproductive role of women gets marketed, there are possibilities of commercialization, increasing role of intermediaries, the health issues related bearing the child, the ethical issues of motherhood etc. The child, being forced to give up breast feeding, may lose immunity and trigger psychological and mental health issues, in the post delivery period. The surrogate mothers face a lot of issues during post surrogacy. Medicines to suppress the breast feeding and other post-natal issues, would have side effects for surrogate mothers. No amount of money can compensate such pains and sufferings. Further, the surrogate mother is detached from the child she carried in her uterus. It is too difficult for any woman to completely detach and give up the baby, which was brought into the world borne by her, but it is happening in the name of commercialization of child birth and hiring out of the uterus. It is difficult to make up the mind to disown the child once it is born, but women are compelled to do so. The issues in the family, after delivery of the child, may have to be discussed, backed by empirical evidence.

Scheme of Presentation of Thesis

The entire book is organized into Seven Chapters,

- **Chapter I:** Introduction and the conceptual description, with the statement of the problem and objectives.
- **Chapter II:** Infertility, New Assisted Reproductive Technologies and surrogacy.
- **Chapter III:** Review of Literature.
- **Chapter IV**: Technical Concepts, and Feminist Theories.
- **Chapter V:** Research Methodology and Design.
- **Chapter VI:** Data Analysis and Interpretation.
- **Chapter VII:** Summary of Findings, Recommendations and Conclusion.

Chapter II

Infertility, New Assisted Reproductive Technologies and Surrogacy

The present chapter presents an overview of the issues of infertility, from a feminist perspective and the emerging assisted reproductive technologies particularly the contributions of surrogacy. On the one hand, the ARTs serve as boon to barren women but at what cost? The economic cost and the physical, psychological pain, that the women and the duo face, till they conceive a child or failure of conceiving a child even after continuous treatment and spending huge amount, for want of a child. Such pressures emerge primarily from the society than from the couples because family, relatives and friends keep on enquiring and keen in knowing on the conception. The inability to deliver a child is something natural and technological development addresses such issues. But given the social norms and expectations, married duos are expected to give birth to a child. If they failed, it is a curse for women in the patriarchal context, where the men, with defect, go scot free. Women are forced to undergo medical consultancy in majority of the cases and the households spend huge amount of money for need of children. Medical technology relieves women from the curse of barrenness and helps them to get a child. Technological revolution plays a vital role in the human reproduction such as cloning, organ transplantation, test tube baby, surrogacy and designer baby. The scientific reproductive technology creates non- sexual reproduction possible. Though it is a boon for the infertile couples, it opens the channels for exploitation of vulnerable women in the society. There are instances where the liberalization and globalization break the moral ethics but still the society welcomes the technological revolution, especially in human reproduction where the negative impacts often go hidden or not taken seriously.

The World Health Organization defines health as "a state of complete physical, mental and social wellbeing, and not merely the absence of disease or infirmity." But infertility strictly affects the personal, political, cultural, economical and social life of a human being. Infertility could also cause enormous emotional pain. The human race believes that only children provide reason to run their entire life. Compared to male, female faced a lot of problems in their personal and political life. Because of infertility, women face various kinds of violence and harassment from home's cultural rituals. There are many reasons behind infertility such as genetic disorder, male infertility, female infertility, sexual dysfunction, and life style. Each year, the percentage of infertility increases. Globalization, Liberalization and Privatization (LGP) plays a major role, in the trendy life of human cell. Because of this modern life, infertility rate has increased. But thanks to the intervention of science and technology, solution such New Assisted Reproductive Technologies help us to overcome infertility. Women, who faced the experience of infertility, social stigma and exclusion from the society, seek help from the new reproductive technologies. Since social taboos and social pressure play a major role in procreation, the problem can not solved by adoption. Due to these reasons, new assisted reproductive technology has gained popularity worldwide.

Infertility

Infertility is defined as the result of a disease (a cessation, interruption, or disorder of body functions, systems, or organs) of the male or female reproductive tract, which prevents the conception of a child or the ability to carry pregnancy to delivery. The duration of unprotected intercourse, with failure to conceive, should be about 12 months, before an infertility evaluation is undertaken, unless medical history, age, or physical findings dictate earlier evaluation and treatment.

Female Infertility: Possible Causes

Female infertility can be caused by a number of factors, which may be natural or due to the life style. The natural causes may be due to scientific reasons but the life style changes do have an impact on the metabolism, obesity, cholesterol composition, hygienic practices during menstruation and infections in the fallopian tubes, uterus etc. which might cause infertility. The following factors may be listed as causes.

a. Damage to fallopian tubes
b. Hormonal causes
c. Cervical causes
d. Uterine causes
e. Endometriosis
f. Unexplained Infertility

Causes for Male Infertility

Male infertility can be caused by a number of factors, which are curable. But in the patriarchal context, male fertility is made to be less visible as it is perceived to be connected with their masculinity. There are women, who failed to bring their partners for consultation and even if the male was infertile, women deliberately took up treatment to deliver a child, with the help of ARTs. Possible causes for male fertility would be:

a. Varicoceles b. Ejaculation problem
c. Anti-sperm antibodies d. Sperm count or motility problem
d. Unknown problem

Historical Perspective of Infertility

According to Hippocratic text, 'pregnancy is how a woman stays healthy. If she can't, she'll be weakened for life, and we all need to cover her in pessaries, cumin, resins, and pity until she can birth a baby. This is probably (and depressingly) a pretty normal attitude for the time.'

In England, infertility was seen as possibly hereditary (weirdly enough) and sterile women were often used as the butt of bawdy jokes. It was women who came in for the most scrutiny, though. Barough gave a whole host of reasons why women might not be able to conceive 'too fat, too thin, womb too hot, womb too cold, womb too foul, filthy and dry (yuck), womb too weak or straight or short.' But mostly, it was just body-policing. Like Hippocrates before him, Barough said that 'infertile women were to blame for their own problems if they had a foolish and uncomely shape, for your reference that was anybody that didn't have big boobs, butt, belly, and stomach.'

Infertility is quite natural for both men and women. Ancient Indian childless women appeared not to have a good life either. One of the earliest Vedic texts reads: "O woe is the woman who does not carry out the provided role of a mother [birth of sons]. O woe the unmarried, woe the childless, woe the mother of daughters, the widow." Women who could not get pregnant were viewed as "possessed by Nirrti," a particularly ferocious goddess and could be cast away out of the family unit. But in the given social structure, the impotency of men was not taken seriously. The new assisted reproductive technology helps to overcome the impotency of men and women.

Assisted Reproductive Technologies (ARTs)

Assisted Reproductive Technologies (ART) is a group of techniques, which are mainly used for the infertility treatment. If the root of infertility is identified by the physician, they may suggest an accurate treatment. However, sometimes no exact issue is identified and then the infertility is labeled as "mysterious." With mysterious infertility, or when conventional treatments have failed, advanced infertility techniques such as In Vitro Fertilization (IVF), Gamete Intra Fallopian Transfer (GIFT), Zygote Intra Fallopian Transfer (ZIFT), Intra Cytoplasmic Sperm Injection (ICSI), Egg Donor, Sperm Donor and Surrogacy, are tried.

Evolution of ARTs

Assisted Reproductive Technologies (ARTs) are the result of a new scientific revolution and the practice of artificial, non-coital insemination became widespread in the mid-Twentieth Century. In 1785, the first artificial insemination was performed by a Scottish surgeon, John Hunte. In 1866, Montegazza, the Italian physician, considered whether the artificial insemination was possible for the soldiers, going into the battle, to have their sperm frozen earlier. If the soldiers were killed in battle, their wives could use the cryopreserved sperm to produce heirs posthumously. The first well known birth, through Donor Insemination (DI), took place in 1884, with the help of Dr William Pancoast, in Philadelphia. The cryopreservation method of human sperm was path breaking in the early 1950s.

First therapeutic insemination was performed, in London, in 1770, by Dr. John Hunter on a patient, whose husband had hypospadias. The world's first test-tube (IVF) baby, Louise Brown, was born in Oldham, Lancashire, UK, in 1978, under the 'care' of Dr Robert G. Edwards and Dr Patrick Steptoe. In 2010, Dr Edwards won the Nobel Prize in Medicine for this achievement (Wade, 2010).

ARTs Applications in India

The applications of ARTs emerged since 1978. It was on October 3, 1978, the first baby, Kanupriya alias Durga, was born in Calcutta, under the 'care' of a Calcutta-based fertility specialist, Dr Subhash Mukherjee. Since then, the field of assisted reproduction has developed rapidly. But due to the lack of peer review and scientific documentation, Dr Mukherjee could not be credited with the achievement. Newer techniques, modifications of existing ones, and new approaches characterize this specialization (Bharadwaj, A. (2016).

Different kinds of ART Techniques

Based on the infertility problem of the couple, the physician advised the couple to undergo various kinds of techniques like *In Vitro* Fertilization (IVF), Gamete Intra Fallopian Transfer (GIFT), Zygote Intra Fallopian Transfer (ZIFT), Intra Cytoplasmic Sperm Injection (ICSI), Donor eggs, Sperm donor, and Surrogacy.

In Vitro Fertilization (IVF)

IVF is a very famous method in ART. *In Vitro* Fertilization (IVF) is a procedure in which egg (ova) is harvested from a woman's ovary and sperm is collected from a man. Then, the ova are artificially fertilized, with a sperm under a laboratory procedure and few months later, the fertilized egg is transferred to the woman's uterus, by injection. IVF method is also called the test tube baby method.

Gamete Intra Fallopian Transfer (GIFT)

GIFT method is not very popular. GIFT is performed where the woman's egg (ovum) and man's sperms are joined in a lab. The eggs are then inserted into fallopian tubes by needle. In GIFT treatments, the fertilization takes place inside a woman's body and not inside the lab. GIFT is a good option when the couple want a more natural ART treatment because it allows the fertilization to take place inside the woman's body naturally.

Zygote Intra Fallopian Transfer

ZIFT method is also very similar to GIFT. The eggs are combined with sperm in a lab and then the mixture is placed back into the woman's fallopian tubes. In the case of eggs are not inserted back into woman's body until the eggs have already been fertilized. As with GIFT, ZIFT process was very popular but because of the success rate of IVF, there are less users of ZIFT.

Intra Cytoplasmic Sperm Injection (ICSI)

ICSI method is most commonly used for duos, with serious problem in man's sperm. In ICSI method, a sperm is placed in the mature egg and after that, the embryo is implanted in the uterus or fallopian tube of the woman.

Surrogacy

Surrogacy is a process in which a woman carries a baby, that is produced from another couple's egg and sperm or donor egg and sperm. The surrogate is to act simply as a carrier till the baby is born. The present study mainly focuses on Commercial Surrogacy, which comes under trade and lot of issues associated with surrogacy.

There are a number of questions to be discussed on the reproductive role of women. Does being a 'mother,' necessarily include the whole gamut of actions, like conceiving, carrying, bearing and rearing a child? What if a woman's womb is removed? or if she cannot conceive? or if she can never give birth to a baby? Does that make her less of a woman? or less of a nurture? or indeed, a mother? Today, babies can be ordered over email, created in Petri dishes (glass) from frozen genetic material, and grown in wombs that are considered to be nothing more than commercial vessels. Today, human eggs are traded like any other commodity and fertile women sell their eggs to sterile women, for the creation of babies, to whom they are not genetically related. Humans, who are created from the embryos born of anonymous eggs and sperm and grown in the wombs of surrogate mothers, will never know their genetic history.

Today, a woman can generate life, which is later grown in another woman's womb and it is nurtured and raised by a third person. So then, does the biological mother cease to be the mother of the child she produced? What about the woman who provided the womb? Does she cease to be a mother once she had given birth? And what is the definition of the third person, who had no biological connection with the baby but only its nurture? Which of these three is the real mother? Who is a mother? and why? Can a father mother a child? Does an embryo have to have a traceable genetic history through its mother? These questions are to be asked, where a baby is not be the product of a simple act of procreation, involving a man and a woman. It is also a world in which the baby makers can play a very important role, in the creation, sustenance and nurturance of life.

Evolution of Surrogacy

In Latin, "Surrogatus" means a substitute i.e. a person appointed to act in the place of another. As per the Black's Law Dictionary, surrogacy means the process of carrying and delivering a child for another person. The New Encyclopedia Britannica defines surrogacy as a practice in which a woman bears a child, for a couple who are unable to produce children in the usual way. According to Warnock Report (1984) HF&E, surrogacy is the practice whereby one woman carries a child for another, with the intention that the child should be handed over after the birth. Surrogacy is the union of science, society, services and the person that makes it a reality. Surrogacy leads to a win-win situation for both the infertile couple and the surrogate mother. The infertile couple is able to fulfill their most important desire and the surrogate mother receives suitable pecuniary reward.

Types of surrogacy

There are two types of surrogacy practices, prevailing in India such as:

Commercial Surrogacy: Under commercial surrogacy, the surrogate mother is paid over and above the necessary medical expenses, by the intended parents.

Altruistic Surrogacy: Under altruistic surrogacy, the surrogate mother receives no financial rewards, for her pregnancy or the relinquishment of the child, except necessary medical expenses.

Commercial Surrogacy in India

Since the onset of globalization, India witnessed massive growth in surrogacy. This happened because of two foremost important factors combination of low medical costs and western-educated doctors, which has been advertised "first-world treatment for third-world prices" (Nussbaum 1998).

Individuals, from the middle upper classes, originate from countries whose surrogacy legislation is either highly restrictive or banned altogether.

Commercial surrogacy has been legal in India since 2002, but India's surrogacy boom began in January 2004, with a grandmother delivering her daughter's twins during 2013 (Sharma, 2012). While many Indian ART clinics are hesitant to advertise the fact that they offer surrogacy services to foreign clientele, an analysis of Indian clinics' websites indicated that "reproductive tourism comprises a substantial fraction of India's assisted reproduction technology clinics' business focus, clustering around its most tourist-friendly locales and surrogacy may be a strong motivator for international clientele" (Deonandan, 2012). India's structure, for commercial surrogacy, is unusual in that there are no national regulations. Compared to other developing and developed countries, India is seen as a haven for global surrogacy. In 2002, India became the first country to explicitly allow surrogacy. The Government of India provides tax breaks to hospitals, treating worldwide patients. India has state of the art medical amenities and technical capabilities, pooled with lower costs than fertility tourists may experience in their domicile country.' The Indian Council of Medical Research estimated that surrogacy is a *$450* million a year industry in India. In 2009, around 1,500 pregnancies were attempted, by using surrogates, at these clinics. The total surrogacy fee ranged from $25,000 to $30,000 and compared to United States, it was very low. From that amount, the surrogate was paid a percentage of the entire cost. The payment, for the surrogate, is estimated to be from $2,500 to $7,000 (Bhalla, N., &Thapliyal, M. (2013). Compared with other countries the total cost of surrogacy was very low in India and no wonder India become famous for fertility tourism.

Surrogacy and Fertility Tourism

Commercial surrogacy gains an additional dimension when it became international surrogacy. An outcome of commercial surrogacy is the emergence of international surrogacy, which has emerged significantly since the first American surrogacy in 1986 (Mechanick, Braverman, & Corson, 1992; Twine, 2011). It is an outcome of medical tourism, to take advantage of the low cost surrogacy, offered in countries like India, which are relatively cheaper compared to other countries. Such methods of moving to other countries to get the surrogate mothers hired, are called \ International Surrogacy, in which commercial surrogacy often occurs across country borders, typically with prospective parents from wealthier countries and surrogate mothers from poorer countries (Humbyrd, 2009). Heightened concern about the exploitation of vulnerable women, is expressed in relation to international surrogacy practices (Twine, 2011). More recently, international surrogacy is also being called Reproductive Tourism (Hanafin, 2006). The fastest developing technique of cross border reproductive care is international surrogacy. It has gained popularity in recent years, due to both international restrictions and to advances in technology, allowing for easier surrogacy (Mohapatra, 2012).

Fertility Tourism Market

Before globalization, when the concept of surrogacy emerged, people thought that reproductive technologies will destroy the culture and believed that it was a curse and practice against gods. Then the thought slightly changed and few people, who were economically wealthy, adopted the technology. Later people became more aware of technology and it was adopted by middle class families too, who are still perceived as the guardian of culture, values etc. Thanks to the science and technology development as well as the people's awareness, the new reproductive technology is welcomed, as one of the strategies to protect women from social taboos and violence. While recipients gained surrogates faced several kinds of exploitation, in terms of payment delayed, not getting the actual payment agreed for, failure to get post natal care etc. It is true that Fertility Tourism facilitates cross-border reproductive care. The Reproductive Tourism or Fertility Tourism is the phenomenon of people, crossing international borders, to use new reproductive technologies. Childless duos are travelling to fulfill their desire for a child. Surrogacy is a multi-billion dollar global industry, raising unique legal and ethical issues and challenges. Smerdon (2008) estimated that from renting of womb, such as creating and delivering baby to peoples, approximately $US6 billion is generated annually. Globally speaking, India is fast becoming the global champion, in providing medical facilities like ART, surrogacy etc.

Legal Surrogacy Cases and the Necessity for Legal Binding and Regulation of Surrogacy Market

There are some cases, where surrogacy ran into trouble due to violation of the rules agreed upon. It occurred in various parts of the globe and a few of them are presented as follows.

Surrogacy violation case of New Jersey

A couple chose not to have baby because of the spouse's sickness. Rather than imagining kids the common way, the spouse entered a surrogacy concurrence with another woman. The man donated his sperm to the surrogate requesting to deliver the kid. But the arrangement was violated, and, ultimately, the case went to the New Jersey Supreme Court.

Surrogacy violation of Japan

Baby Manji is a child, delivered by an Indian surrogate. Her authorizing guardians were a couple from Japan, who filed for separation, in the blink of an eye before the child was conceived. The father, as yet needing to deal with the baby, confronted serious legal issues as the Indian law disallowed single men to receive. Neither the proposed mother nor the surrogated mother wanted to take care of the child, Manji. The child was in the long run allowed to leave for Japan, after the Japanese Government provided a one-year visa to her on compassionate grounds. The child's grandmother was given authority over the infant. As a consequence of this case, the level headed discussion inside India, about surrogacy, has escalated. In the debate that followed this incident, in the absence of a legitimate contract between the parties, paved the way for potential misuse of local surrogates.

In the case of the Japanese parents, who failed to claim the baby due to their separation, the child became a victim and the technology could hardly protect the child. Hence the need for strong regularity framework.

These issues exist since surrogacy contracts are not clear and hold no legal sanction. A few nations passed surrogacy enactment. Those countries which did have these laws, regularly neglected to enforce them. It is presumed that restorative innovation, particularly conceptive innovation, required no defense. But with the commercialization of surrogacy, social, demographic, moral, lawful and philosophical issues have been raised. As the informed discussions surfaced, these advancements can modify the very soul of human progress. It may achieve the rebuilding of society on the lines of a 'conceptive house of ill-repute model' in which women can offer regenerative favors the same way old fashioned whores sold sexual ones' (Ravindra, 1992). Because of the issue, the Indian Council of Medical Research in India had drafted ART Bill Regulations 2008 and a revised new bill on 2016, for regulating commercial surrogacy, especially for foreigners.

Surrogacy: Legal Background an International Perspective

Surrogacy is legal in some cases whereas in others, it is illegal. Some of the parents have adopted the procedure of surrogacy. Every country is having a bit of difference, in the laws for surrogacy, which are presented.

Status of Surrogacy in some other countries

S. No	Country	Status of Surrogacy
1	Australia	Prohibits commercial surrogacy. The state of Queensland bans all forms of surrogacy. in the in the other Australian states such as Victoria, the Australian capital territory, Tasmania and south Australia commercial surrogacy is allowed
2	Canada	Prohibits commercial surrogacy. is silent on non commercial surrogacy
3	China	Prohibits commercial surrogacy

S. No	Country	Status of Surrogacy
4	France	Prohibits all surrogacy
5	Germany	Prohibits all surrogacy
6	Italy	Prohibits all surrogacy
7	Israel	Prohibits commercial surrogacy but allows non commercial surrogacy
8	India	Permits all surrogacy (Recently Govt. declare to ban commercial surrogacy)
9	Norway	Prohibits all surrogacy
10	Russia	Permits all surrogacy
11	Singapore	Prohibits all surrogacy
12	Sweden	Prohibits all surrogacy
13	Ukraine	Permits all surrogacy
14	USA	Commercial surrogacy is not allowed in some of the states. States that allow but regulate surrogacy are: California, Arkansas, Florida, Illinois, Nevada, New Hampshire, Texas, Utah and Virginia
15	UK	Prohibits commercial surrogacy but allows non commercial surrogacy
16	Vietnam	Prohibits all surrogacy

(Source: SAMA Research Group, 2012).

Based on the experience of the above countries recently India prohibited the commercial surrogacy because some people misused the illiteracy of poor women and did not fulfill the contract. As a result, death of children and surrogate mothers was increasing day by day. Also, practicing surrogacy was deemed unethical in Indian culture and hence the Government of India banned commercial surrogacy.

Proposed New Draft of Assisted Reproductive Technology Draft Bill – 2016

The state of surrogacy permission and legal sanctions, regulations in various countries across the globe, revealed that there are differences in their provisions and legal safeguards. Only in few countries like Israel, Iceland, surrogacy is possible without much restriction. In many countries it is completely banned. It is good that in some of the countries, it is strictly prohibited to have any form of surrogacy and at the same time alternatives have also been provided towards adoption etc, particularly in New York. In the Indian context, it was open without any restriction and the cross border states really provided greater scope for reproductive tourism but later due to the legal issues, particularly violating the norms, the civil society fought for regulating the same and hence the New ART draft bill was enacted and it has restricted the free access to such surrogate trade by foreign couples, particularly after the case Japanese couple case of disowning after the conception.

Science and technology are quite supporting the human lives in several ways. Reproductive technology has grown to the extent of prompting a woman to hire out her uterus, to bear a child and help an infertile couple, to have a biological child. Infertile women are put to extreme stress and abuse in the patriarchal cultural context. Assisted Reproductive Technology has really become a boon to such childless couple. Surrogacy is one form of ART, which helps the childless duos, through the mean of external duos, by carrying their foetus, both as service and for pecuniary motive. There has been a huge commercial market, particularly for the foreigners, who come and stay to get a child which is relatively costly in their own countries. It is a highly exploitative market, where the mediators get benefitted to a large extent by taking advantage of the desperate requirement of the child by the childless couples. There are a number of issues, faced by the surrogate mothers because their rights are not protected. The dynamics of the market for surrogacy really push the surrogates to extreme vulnerability, where the payment fixed at the time of agreement is often not paid. The amount agreed is not actually paid in time or paid less than what was being agreed upon. Surrogates from economically backward households, are forced to convince the family and relatives, they lack post-maternal care particularly the kind of medicines consumed, must forego breast feeding etc and these issues seriously cause several physical and health problems. Above all, surrogates do not have anyone to report over the above issues. On the demand side, the mediators, the infertility clinics, the agents or brokers do not maintain transparency and they do not allow, direct contact between surrogate and the commissioning couples. This lack of transparency makes the surrogate mothers to suffer. There are several other issues in the surrogacy market dynamics, where each agent would wash off their hands and surrogate mothers cannot regularize their rights. As a result, the Indian Government has recently banned such practices for foreigners. The recent draft ART bill aspires to safeguard women from victimization and exploitation, from present commercial surrogacy ecosystem.

Remarkable Features of New Draft ART Bill -2016

The following are some of the provisions of proposed surrogacy draft Bill of 2016,

- ✓ Single parents, divorcees, unmarried couples, living together duos, lesbians or gay couples are barred from opting for surrogacy.
- ✓ Married women alone can practice surrogacy.

- ✓ Foreigners, Non Residential Indians (NRIs) or persons of Indian origin (PIOs) who hold Overseas Citizens of India (OCI), cannot look for surrogacy.
- ✓ A couple, with already a surrogate child, cannot opt for surrogacy for the second child.
- ✓ Couples, who already have adopted a child, are also barred from surrogacy.
- ✓ The surrogate mother should be married and have a healthy child
- ✓ A woman can only become a surrogate mother only once in her lifetime.
- ✓ Only 'altruistic' surrogacy will be allowed, in which case married couples and close relatives alone can become their surrogates without any money exchange, except for medical expenditure.

To sum up, the above specific features do nothing to alleviate the vulnerability of working class women in India and its only its further leads to multiple exploitations. On the other hand, the New Assisted Reproductive Technologies made non- sexual reproduction possible in the technological era where one can give birth even though she was never touched by a man. Baby Makers, Arati Prasad, explored the possibility of there being an ultimate solo parent in the future. This woman would use her own stem cell and an artificial Y chromosome, to produce healthy new eggs and sperm at any age. She would be capable of reproducing; all by herself, by making one of her eggs behave like a 'pseudo sperm.' The embryo could be gestated in an artificial womb would this artificial womb, therefore, become a great biological and social equalizer? And if it happened, what will happen? It's beyond belief.

Chapter III

Review of Literature

This Chapter presents an overview of Surrogacy, as an Assisted Reproductive Technology in the first section and review of related literature in the second section. Review of related studies would provide the status of research in the field and inform the gap existing, to proceed with the current research.

Section I: - Overview of Surrogacy as an Assisted Reproductive Technology

Surrogacy is sought after by the commissioning parents, who wish to have their biological child, due the inability of the women to bear. There are various kinds of arguments raised on Assisted Reproductive Technologies because of the issues and challenges related to post surrogacy issues. The differentiation between altruistic and commercial surrogacy is also a dominant theme in the literature.

Commercial surrogacy is more controversial than altruistic surrogacy as it raises number of questions on commoditization of women, children and reproduction in general. According to, Liberal feminism, "reproductive technologies liberate the women from the reproduction." But Marxist, Socialist and Radical feminism were oppose the technology. These feminist strongly criticize that surrogacy commoditizes women's body and reproductive rights, creates division of labour, gender discrimination from the society. The natural question's which would arise are: Can't the women make a choice of not to reproduce and resort to adoption? Will surrogacy not pushing the women to treat her body as an object?

While opponents of surrogacy demand banning surrogacy, others hold the view that it should be regularized. It is quite likely that if the surrogacy were to be banned, black market will emerge; it will create huge demand for surrogate mother. While, banning of commercial surrogacy leads to more exploitation of poverty and illiterate women, this would cause more harm than good. Laws are non-existent or poorly enforced. Surrogacy creates multi angle questions to which the response of a Researcher in Women's Studies would be guarded. Feminists have differences on the issue and challenges of ART. Based on the issues and challenges, various kinds of arguments have emerged and feminists such as Liberal, Marxist, Sociologist, Radical, Postmodernist feminists critique ARTs on the grounds that it is trained with patriarchy, and capitalism subordinates and suppresses women in the society. The following section presents the research studies, on various aspects of surrogacy, which would also present the issues related to surrogacy.

Section II: - Review of Literature

The review of related studies is presented, based on various themes on which the researches have been done. They are presented as follows, where majority of the research have been found outside India.

Empirical Studies

Related researches on ARTs and Surrogacy have qualitative, descriptive, narrative and empirical focus.

Amrita Pande (2009) interviewed 42 gestational surrogates, from Akanksha Infertility Clinic, their husbands and in-laws, and the director of the clinic, Dr. Nayna Patel, between 2006 and 2008. According to her reports, relatives were free to visit surrogates, but due to the costs of travel, many surrogates did not see their families while they were pregnant. Surrogates reported missing their children. Some also reported that not having to take care of their household or do other works, was a welcome break. The payments, that surrogates received for carrying a baby, often equaled four or five times their annual household income.

Appleton (2004) found that majority of surrogate mothers were married or had partners. Some surrogate mothers, in this study, justified their decisions on altruistic grounds. It is the married women, who extended their support, to become surrogates which is universal also.

According to Baslington (2009), the majority (ten out of fourteen) of surrogates reported that they were able to cope well with the relinquishment, though the general experience was to feel unhappy in the short term and if a good relationship was forged with the commissioning couple then this also eased the burden of relinquishment. From the above statement, it is evident that the surrogate expected relationship with the commissioning duos.

Chaves (2011) reported that usual pregnancy risks apply to surrogates also but they are accentuated by the effects of ART, resulting in multiple births or there would be need for selective reduction abortion. There have been cases of surrogate mothers dying because of pregnancy complications, something that accentuates the issue of postpartum risks. In the case of traditional surrogacy, procedures such as collecting an egg carry a certain risk and might involve physical and psychological suffering

Ciccarelli (1997) and Ragone (1994) found that American surrogate mothers had different levels of education and they were in their late 20s and early 30s years of age. They entered into the surrogacy by choice. The commissioning parents were generally older than the surrogate mothers and they were more highly educated and had high incomes. When the surrogate mothers consented to surrogacy by choice, it could be questioned either on moral or monetary ground.

According to **Ciccarelli (1997),** fourteen surrogate mothers were asked to report their feelings or concerns about relinquishing the child. One mother reported emotional distress over the relinquishment and two others reported a strong instinctual urge, to bond with the child. The remaining eleven did not feel bonded with the child, which may seem to indicate that for the majority of surrogates, the issue of having to relinquish the child did not appear to be a problem. However, surrogates had to employ cognitive dissonance reduction strategies to cope with their loss, implying that the issue was deeper, and depended upon the strength of the bond between gestational mother and child.

Dixit (1994) maintained that there are at least two roles for economists in the policy-making process. One is to provide information and another is to attempt to shift the frame of reference in which policy decisions are taken. Both these roles are appropriate in the case of surrogacy, as evident in new surrogacy bill in the Indian context.

According to **Iowa Institute (2012)**, lack of regulation raises increasing concerns, because surrogates are often destitute and illiterate. They are kept enclosed in clinics, they do not enjoy counseling or legal services, they are subjected to decisions taken by the doctors involved and they also undergo tests and operations, including often unnecessary caesarean sections, for quick delivery. Iowa Institute (2012) also found that the aftercare of the surrogates is an important issue, particularly in developing countries. Long-term harmful effects caused by fertility drugs, and surgery-related complications are frequent and might be more acceptable in a developing country context. Serious conditions like cancer or sterility can also be related to surrogacy. The question remains as to who bears the responsibility for treating health problems, resulting from surrogacy pregnancies? This is really an unresolved question and in the absence of post surrogacy treatment or counseling, surrogate mothers may be affected by health issues.

Javada et al. (2003) found surrogate mothers repeatedly reported a sense of self-worth, resulting from the experience of surrogacy. Majority of surrogates were open about their surrogacy experience with family and friends, even though some family members and friends reacted negatively in the beginning. In the end, family and friends frequently accepted the surrogacy arrangement, and felt proud of the surrogate mother. Surrogate mother's partner was mainly supportive.

Parker (1983) interviewed 125 applicants to a surrogacy program. Dismissing the supposition that surrogates were driven by financial interest, he found that wanting to help others, earning money at home, and enjoying pregnancy, complemented one another. In an especially interesting passage, Parker also notes that a subset of women (he makes no mention of how many) "wanted to be pregnant for the rest of their lives and they felt more content, complete, special and often felt an inner glow."

Parker (1983) and Ragone (1994) both studied the motivation of gestational surrogate mothers. Majority of these women were from the working class, with low educational backgrounds and almost all expected to be paid for their service. Although financial compensation was stated clearly as an important element in the decision of the women in Ragone's study, her hypothesis is that the surrogate mothers see surrogacy as a vocation, a calling, a means for expressing and fulfilling themselves. In the West, it could be observed that the surrogacy has been taken up by choice and not just for marketing and earning money.

Ragone (1994) emphasizes social and cultural factors, in her explanation of what motivates women to become surrogates. The pregnancy allows the woman to work at home. She spends the cash on her husband and children, to compensate them for the inconvenience to which her atypical pregnancy puts them. Although carrying a baby for someone else, violates conventions about child bearing, surrogates represent their actions as culturally appropriate, falling within the traditional roles of the nurturing woman. The surrogate creates motherhood and fatherhood, for the intended parents, by giving them the gift of a child.

Ragone (1996), in a study, says that her informant had three motives. They wanted to help infertile couples, they wanted to earn money at home, and that they loved being pregnant. Ragone (1996:360,363) also found that, after studying six surrogacy programs, "Without exception, when surrogates are asked whether they think of the child as their own, they say they do not." Women, who are concerned about a biological link to the baby, choose gestational surrogacy despite the higher risk of medical complication involved in the process of transferring and implanting eggs. They typically hand the babies over without incident and it is clear that, regardless of the physiological relationships, bonding does not constitute an impediment to surrogate motherhood.

According to **Van Zyl and Van Nierkerk (2000**), the typical responses, given by surrogate mothers, with regard to how they felt about their relationship with the foetus, were as follows, "I don't think of the baby as mine. I donated an egg I wasn't going to be using;" "The baby isn't mine. I am only carrying the baby;" and "I am strictly a hotel." That must be the perspective of surrogate mothers so that they may not suffer with bonding and will not end up with psychological issues.

Studies on outlook against Surrogacy

As pointed out already, there are different schools of feminist thought on surrogacy. One school expressed that it liberated the women and others underlined that it objectified women.

Catherine London (2012) held that historically, women of color were subject to forcible sterilization and oppressive reproductive restrictions. Black slave women were often classified as "breeders," who had no legal claims to their children. Also, gestational surrogacy is another instance where Black women are valued for their effective labor, in the service of White people.

Corea et al and, Arditti et al. (1987, 1989) found that the new technologies are scientifically-based, medicalized form of violence against women's bodies. As such, they are to be feared and resisted as patriarchal practices, which continue the long-standing exploitation of women's bodies for masculine advantage.

Dorothy Roberts (2008) informed that some agencies already have directories that "display photographs of and vital information (height, hair color, racial origins) about women willing to be hired to gestate a baby."

Dworkin, Margaret Atwood, Gena Corea and Robyn Rowland emphasized thatre productive technology posed enormous threats, to whatever powers women still possess and that biological motherhood ought not to be forsaken in favor of artificial motherhood. The replacement of women's child bearing capacity by men-controlled technology, would definitely remove women's biological burden but it would also leave women without a product with which to bargain.

According to **Elizabeth Anderson (1990),** surrogacy contracts alienate a woman from her love for the child and frequently involve exploitation, as surrogate sellers have less wealth and they are more emotionally vulnerable than buyers. Other objections stress the weakening of the link between parent and child, and the special vulnerability of children.

According to **Golombok et al., (2006),** compared to natural conception families, the surrogate mothers showed more positive parent-child relationships (higher levels of joy and competence, and lower levels of anger and guilt) than mothers with a naturally conceived child, and the surrogacy fathers reported lower levels of parenting stress than their natural conception counterparts. Regarding the children, no group differences were found for either cognitive development or psychological adjustment.

Gupta (2012) argued that transnational surrogacy in India, reflects many of the inequities in India, as elsewhere, like disparities in gender, race, class, and nation place, in some women's reproductive projects.

According to **MacKinnon. A (1995),** opponents of surrogacy successfully shaped the social meaning of surrogacy, as a degrading business in which poor women were unfairly exploited and coerced by profit seeking brokers to sell their babies. Hence, there was almost uniformly hostile stand on surrogacy and favorable to restriction.

According to **Margaret Jane Radin (1988) and Carole Pateman (1983),** the labor of bearing a child is more intimately bound up with a woman's identity than other types of labor. Contract pregnancy involves an alienation of aspects of the self as to make it an illegitimate practice. The work of pregnancy is long-term, complex and involves an emotional and physical bonding between mother and foetus.

Piercy and Firestone (1994) were severely criticized by Marry O'Brien and Adrienne Rich. O'Brien analyzed reproduction through the lens of male alienation in reproduction. Patriarchy is men's compensation for and attempts to counter act the alienation from reproduction.

Satzs (1995) held that commercial surrogacy allows women's labor to be used and controlled by others and reinforces stereotypes about women. For example, pregnancy contracts give buyers substantial control rights over women's bodies and rights to determine what women should eat, drink and do. They also may deepen the stereotype of women being baby-machines. Finally, the race and class dimensions of such markets also need to be considered.

The studies, reviewed under commercial surrogacy, revealed that the rights of surrogate mothers are violated in several ways and they are just treated as baby producing machines as there is no connection between the mother who bore the child and the child post surrogacy. This is the critique of feminists where the women are just treated as objects and not as human beings. Added to the economic issues, the women surrender themselves to being almost a product and sell their uterus, bearing the physical, mental and psychological consequences. Reducing women to such a state further pushes women to subordination, which was prevalent since the Epics. Women, who were in the serving community, were treated just as a commodity employed only to breast feed the queen's child. In the corporate language, it is a blessing for women where the market offers an opportunity for the poor and economically vulnerable women and who do not really question the injustices, if any, particularly violation of the rights of surrogates.

Women's Reproductive Role and its Exploitation

There are a number of studies on how women's reproductive capacity is exploited.

Davis (1993) stated that during times of slavery, slave women were often used as birth or genetic mothers and like the surrogate mothers in the present context, they also possessed no legal rights as mothers. In the light of the commoditisation of children, they have the same status as surrogate mothers have in contemporary times. Another similarity is that slave mothers could not speak freely about their pregnancy and the children they carried and it is an aspect that is also present in surrogacy as a result of social stigma. Davis is worried that given this history, poor women may be transformed into a special caste of hired pregnancy carriers. She believes that with the commoditisation of labour services of pregnant surrogate mothers, money is being made, which implies that someone is being exploited. Davis continues by saying that surrogacy appears as a generative procedure of life, and what is really generated seems to be sexism and profits. Surrogacy is not a new phenomenon but it had been practised earlier in the primitive and slavery era.

Horsburgh (1993) believes that surrogates are physically exploited once they have signed contracts agreeing to give birth to babies for clients. If there is a reason to abort the foetus, because of medical reasons or client's demands, the surrogate mother must comply. To make matters worse, if the pregnancy is indeed aborted, the surrogates often receive just a fraction of the original payment.

According to **Kembrell (1988),** the practice of surrogacy exploits women economically, emotionally and physically. An important factor is that most women, who get involved as surrogates, do so because they are in desperate need to meet economic challenge to maintain their family. In addition, agents are involved and arrange contracts of questionable legality. Those contracts require women to undergo all the rigors of childbearing, and eventually they have to give the child away. The surrogate mothers are often unaware of their legal rights and due to their financial situation, they cannot afford the services of attorneys. Once the surrogate mother has signed the contract, it is impossible for them to escape. He further mentioned that the practice of surrogacy represents a new and unique form of slavery of women.

According to **Rapp (1988),** Feminism challenges the new reproductive technologies, as if there is a unified category called 'woman,' whose natural ability to bear children now stands threatened by the total male, mechanical medical take-over. Labeling a single oppositional stance as 'feminist' and anything else as 'not,' prematurely forecloses the strategies we need to develop.

Psychological Perspectives

There are several studies on the emotional perspective of how surrogate mothers suffer right from accepting to serve as surrogate, till they get relieved, how they suffer in the post surrogacy in terms of pressures from all possible angles and finally, parting with the very child borne by her, without being considered in the post delivery and just being thrown out. In certain cases, the commissioning parents allow the surrogate mothers to see the child and also continue the relationship but legally, there is no such arrangement to connect with the commissioning parents. Even if the surrogate mothers do not show any interest to see the child back, it is because they have made up their mind not to claim ownership anymore and the kind of detachment developed, really contributed to giving up any emotional imbalance post delivery.

According to the **Centre for Social Research (2012),** research, on surrogates, has demonstrated preoccupations related to leaving their children during their stay in sheltered accommodation for nine months, together with exhaustion and considerable pain after each IVF treatment.

Foster (1987) underlined that many surrogate mothers faced emotional problems, after having to relinquish the child. She recalls a woman, who said that she started praying not to go in labour so that she and her child could stay together. However, other authors disagree with Foster.

Golombok et al., (2004) asserted that the impact on the surrogate's psychological health and on relationships with members of her own family, in the long-term, is also unknown. Experiences may also be different for genetic surrogates (those who use their own oocyte for the surrogacy), compared with gestational surrogates (those who use the intended parents' gametes or donor gametes) or between known surrogates (family members and friends of the intended parents) and unknown surrogates (those who met the intended parents for the purpose of the surrogacy). By nature and bonding etc, the genetic surrogate might get affected more than the other category as it is genetically connected.

According to **Jadva et al. (2003),** neglected psychological dimensions involve the bonding of the surrogate mother with the child, which might be linked with feelings of guilt when the child is given away and which can have impact on the psychological wellbeing or mental health of the surrogate.

Jadva, Murray, Lycett, MacCallum and Golombok (2003) showed that surrogate mothers did not appear to experience psychological problems as a result of the surrogacy arrangements. Although they acknowledged that some women did experience emotional problems in handing over the baby or as a result of the reactions around them, these feelings appeared to lessen during the weeks following the birth.

Ragone (1994) found that when therapy, designed to maintain the "desired state of mind" of the surrogate was withdrawn, due to the surrogacy agency becoming bankrupt, all the surrogates involved subsequently expressed intense separation anxiety. Ragone states: 'When the support services are removed and the structure of the program dissolves, it is difficult, if not impossible, to maintain the prescribed and desired boundaries between the surrogate and her child and hence surrogates report feelings of loss, pain, and despair while parting with the child.'

Susan Golombok (2011), examined the impact of surrogacy on mother child relationships and children's psychological adjustment. Assessments of maternal positivity, maternal negativity, mother-child interaction and child adjustment, were analyzed, with the help of 32surrogacy, 32 egg donation and

54 natural conception families, with a 7-year-old child. No differences were found for maternal negativity, maternal positivity or child adjustment, although the surrogacy and egg donation families showed less positive mother-child interaction than the natural conception families. From this study finding function of egg donation and surrogacy families were healthy in the early school years.

Tieu (2009), documented that important biological bonds are developed during pregnancy. The aroma of an infant is attractive to the mother while sight and skin to skin contact further promote psychological and physiological bonding, as important hormones like oxytocin are in operation. Surrogacy interrupts the process of bonding that starts during gestation and continues after birth and this is a very important reason why many surrogates refuse to relinquish the child.

Commoditisation and Objectification of Women's Body

It is often pointed in media that women are objectified in the advertisement, where the women's body parts are being exposed to capture the market. Same logic applies to surrogate market also.

Berkhout (2008) held that in the case of commercial surrogacy, surrogate mothers are treated as an instrument as tools for the commissioning genetic parents to obtain a child that is related to them. Similarly, Rothman (1989) argued that the technologies re-essentialize women by reinforcing patriarchal roles and objectifying women's reproductive potential. It is true that the women are reduced to mere product as argued by the researchers.

Centre for Social Research (2012), in its research, has shown that delays in signing a contract entail dangers of exploitation by doctors, clinics, or intended parents. Indian surrogates attracted by an offer $6000, for making up the surrogate contracts, are led to their decision because of poverty, unemployment or the need to finance the education of their children.

Radin (1996) raises the issue of surrogacy and calls it baby selling. She states: 'if it were okay to think of children as property, then it would be okay to buy and sell them and if it is not done to buy and sell them, then maybe it's not done to think of children as property. So also the surrogate mothers need not be considered as property.'

According to **Rimm (2009),** the concept of "renting out" one's womb is unnerving and it is often viewed as a form of prostitution, organ selling, or slavery, as each of these activities involves the payment for use of someone's body or parts of someone's body for another person's benefit. This is the extreme view point where the initial resistance to accept to become surrogate, got internalized to, by every woman in the Indian context, and hiring out uterus, and agreeing to bear the other man's child in one's uterus, is almost equal to prostitution. But in the market economy context, it should be seen as a transaction where agreements are being made between buyers and sellers and the good is delivered. So market will reduce the woman and the child to mere product and commercial transaction. The society will forego the values, culture etc, in such areas or may justify and accept. Patriarchy will give up its values for the sake of market. The question is the society is so insensitive and do not encourage adoption or other ways of having a child and rather makes use of the technology and exploit the vulnerability of the women from poverty and illiterate families, by thoroughly brain washing through different actors, in the surrogate market. The worst is that there are demands for designer babies, preference for sex of the child, fair color, clan, etc. Mythology has evidences to show as to how other women bear the reproductive needs of the upper class women. The Tamil film' Sindhu Bairavi focuses on such perceptions where the great musician suffers without a child and another woman enters into his life, develops a relationship and delivers a child and leaves, handing over the child. Same has been portrayed in 'Kalki' film also.

Stand Point of Economists

Economics just encourages the market. Market is for profit. It will work when there is prosperity. It does not have any consideration on social concerns or ethics. There have been studies, which focused on the economic perspective. It is based on both demand and supply side. The commissioning parents' interest, to have their own biological child, irrespective of how much it would cost on the demand side, and surrogate mothers pushed by their poverty, willing to supply their uterus, reduce the surrogate market to an exchange. The affluence vs poverty nexus becomes so strong, where a trade-off takes place with the members of the family on economic earnings vs support. It is not a surprise where the price is being paid both the sides for infertility and the social expectations that only biological child should be owned. The price, that the commissioning parents pay, is mere economic price, but what the surrogate mother faces is a social, cultural, physical and emotional cost, which cannot be replaced by anything where women are almost forced to accept converting their heart into a stone. It is not a big issue that market will accept the women's body to be sold (uterus), to bear another man's child, but such focus is not found in other contexts, where women's relationship with men or lower caste men etc, are being treated as a sin and violation of culture. Technology, with capitalism, will act as perfect partners through market, to make women to be a commodity to be traded.

According to **Janice G. Raymond (1989),** the commodification of women's bodies in surrogacy and the new reproductive technologies in general, is a gross violation of women's dignity. Women are increasingly treated as reproductive environments and receptacles utilized in medical science as matter to be experimented on, without bodily integrity - in short, without dignity.

Judge Posner (1987) is of the view that efficiency will improve with free trade. This will happen when there are parents, who are eager for children and women who are anxious to be surrogates. However, once this trade of parental rights is prohibited, black markets will come into existence. It is in this context the recent Surrogate Bill also is being critiqued. When it is controlled in the name of providing rights, black market will emerge where what is being referred to in economic theory, as 'Moral suasion' 'free rider' of Rawl's (2017) concept will work, where the ethics will be paid as the price for the market. More economically vulnerable women, may be offering themselves as surrogates still with lower payment. Doctors and other stakeholders may be pressurized to maintain transparency and further exploit the women.

Kopytoff (1986) held that commoditization, commodification and commercialization contain undertones of what Kopytoff has called a "moral economy." He noted specifically when human body parts became commodified, in the case of blood, organ, ova and sperm, along with surrogacy, the fine line between persons and things became hardly debated. It needs to be debated in a feminist perspective.

Roberts (1995) and Field (1992) asserted that through the form of relationship that the surrogates develop, from kinship ties that disturb the sanctity of biology and genes within a system, that might well be the pinnacle of the commoditization of the genetic tie, presents another side of the surrogacy contract. The biology does get affected by bearing a baby for another woman which cannot be compensated by money alone.

Ryan (2009) held that sperm, ova, embryos, and wombs etc are all desirable and profitable commodities, the trade off which serves the intersecting interests of many parties (donors, recipients, infertility specialists, and IVF brokers) and it is facilitated by advances in global communications. The trade off has been quite in contradiction with culture, which is not the concern of the technology and market. The question is, why does the society accept such changes and justifies it on the grounds of want of a child while it is so strict on other aspects where women are expected to be protected. It would not be surprising in future that women will be put into technological and market exploitation on beauty, appearance etc, where the plastic surgery market is already emerging.

Thompson's (2005) study on assisted reproductive technologies in America, suggests that commodification leads to a different kind of capitalism. She compares and contrasts the "Capitalist Modes of Production" with "Biomedical Modes of Reproduction."

Tieu (2009) underlined that one must depersonalize the whole process and treat the surrogate as an object and commodity, or a 'womb for rent,' in order to make the breaking of the biological bond more palatable but this entails making a surrogate susceptible to exploitation and coercion.

Watson (1992) categorically emphasized that by creating an economic market for surrogacy services, women who serve as surrogates, will become targeted for their "race, beauty, health, or intelligence."

Discrimination: Sperm Vs Uterus Selling

There is another school of thought which holds the view point that surrogacy is selling sperm and not baby and argued that what is wrong in women hiring out uterus. Why should that be treated equivalent to prostitution or body sale etc.

Michael J. Sandel (1998) stated that coercion and corruption play a major role, because commodification of human body parts is intrinsically degrading, a violation of the sanctity of the human body. Those, who dispute the analogy, argue that commercial surrogacy is more like selling sperm than selling a baby. If it is morally permissible for men to sell their reproductive capacity, this argument goes, why is it not morally permissible for women to sell theirs? There is a moral implication of the analogy between surrogacy and sperm-selling. This analogy is typically invoked in defense of surrogacy. If men should be free to sell their reproductive capacity, the argument goes, shouldn't women also be free to sell theirs? Isn't it unfair, isn't it discriminatory, to allow one but not the other? But the feminist concern is that the women's body is reduced to just commerce towards reproduction and put pressure on women that they should conceive. Surrogacy is sought only as a last resort and women's body is being put into the maximum pain and vulnerability in the name of treatment for infertility. Whether it is the mainstream perspective or the feminist argument, women's body cannot be reduced to a commodity and rather they can remain without child or adopt the child. That should be the perspective, but technology and growing capitalist perspective, capitalize the market for infertility treatment and infertility treatment clinics are mushrooming. A word of caution. Normally, the sperm of the husband alone may not be inserted as the focus is to turn successful and get a baby for the longing parents. So sperm donors market is being utilized for the same and it is not a guarantee that the husband's sperm alone is injected to the egg of the women. Here comes the argument of prostitution. The feminist critiquing focuses equally on both perceptions and argues against women being pressurized to conceive.

Feminist Stand Point

As already pointed out, feminist stand point is not to treat women as mere reproductive machines and exploit their economic backwardness. The mainstream thinking only appreciates the technological advancement and resorts to any method to get the infertility treated, for which the patriarchal based impositions never spare women, free from such thoughts of having one's owns child. In fact more than men, women aspire to have a child by any means. This type of thinking emerges from middle class where it is perceived that marriage or family will be complete only when a child is born. The infertility is a taboo and curse for women, as the women get discriminated and even the society arranges to second marriage for men, for the sake of having a heir for the family, for which even wives had agreed and gave way to get second marriage for husbands. Thanks to technological development, women are spared from such agony of sharing her husband with another woman. Men do not resist such second marriage for the sake of a child. In certain cases, the second wife also would not have conceived and they might have to go for third marriage, where again the result might be negative. The pain with the infertility treatment, now is the recurrent failure of conception and repeated treatment which must be addressed.

Andrea Dworkin (1983), the well-known American feminist, asserted that: 'motherhood is becoming a new branch of female prostitution, with the help of scientists who want access to the womb, for experimentation and power. Feminist concerns cannot be expressed beyond this view point. Both science and capitalism have made the women, to just hire their body, for meeting the economic needs. More than that, the very value and recognition being given to women in Indian cultural context is motherhood and if at all, the women are respected and recognized, is for their reproduction and with the penetration of assisted reproductive technologies, the women's role may still be sub-ordinate and people may resort to.'

Christine Overall (1993), a Canadian feminist ethicist, argued that the informed consent, signed by the surrogate mother, is not truly a free choice and that it is not possible for the surrogate mother to understand fully, the trauma of giving up her child. She also contended that surrogate mothers have limited income and personal security, as well as low levels of education. The relationship is not one of equals but rather, one of exploitation of poor marginalized women. In short surrogacy, as "reproductive prostitution," is more or less comparable to slavery.

Gene Correa (1985) questioned technological intervention in what she considers the otherwise natural process of reproduction. She held that women derive special attributes, power, and meaning from their procreative role, and argued that they will lose their value if reproduction is technology controlled by male medical practitioners. Thus, technology is seen as having the potential to erode true female consciousness and female affinity with nature.

Hatzis (2003) and **Centre for Social Research (2012),** explained that the subjects at the receiving end of health care in the destination country are the surrogate mothers and the baby. Health issues regarding the surrogate mother are of significance, as her preparation and the period after insemination, involve injections of hormones, estrogen and progesterone, taking pills, as well as adopting a particular lifestyle.

According to **Marsha Darling (2006),** the concept of women's bodily integrity is threatened by the extent to which women's biological and reproductive organs, tissues, cells, including ovum and genes, are quickly becoming 'spare parts' in a medical industrial complex. Reproduction is imagined as an industrial process by the biotech industry, and women are sought after, as consumers of the very technologies, that will weaken women's right to bodily integrity. The biotech industrial contribution is still perceived as a blessing for women to have child, whatever the type of pain and the medical industrial focus may be. Such perception is really least recognized by feminists, who acknowledge the reproductive role and procreation by women which nature has given to only women regardless of the very debate whether all women should procreate as there are single women, and LGBQT, who by choice, chose a different relationship or does not want to get married. It is time and again emphasized that women's role is productive, reproductive and community roles but unfortunately, only the reproductive role is being put at the centre stage and all policies, science and technological advancement, medical research etc keep women's reproductive role as a priority, which needs reconsideration.

McDermott (2012), a well-known feminist, condemns 'baby selling,' referring mainly to traditional surrogacy, which involves giving up not only the babies' surrogates carry but also their hereditary material. Given that the hiring out uterus to baby bearing is done, it does not confine to that part of the body but the whole reproductive system, which must be recognized while establishing any technology to women. Glorifying motherhood serves as the base, which commences from the family and gets extended to the civic, industry and community and hence the other roles get sidelined or even disregarded.

Pande (2010), criticized gestational surrogacy as it is being organized and implemented "as the ultimate form of medicalization, commodification and technological colonialization of the female body, and as a form of prostitution and slavery, resulting from the economic and patriarchal exploitation of women."

Shah (2009) held that surrogacy can be seen as 'a possible choice, restricted but made with dignity, knowledge and consent' and warns against the glorification of motherhood. Shah points out that the conflict over surrogacy, surrounds the 'disconcerting use of the language of 'rights' and 'choice,' by the promoters of these businesses, on behalf of women going in for these technologies,' on the one hand, and 'the assertion of rights over the body as a resource,' on the other hand. What remains unaddressed, particularly because of the absence of state regulation in the Indian scenario, are question pertaining to the inefficiency and health risks of invasive procedures, the disruption of women's lives, the insecurity of surrogacy dealings, the possible legal tangles regarding the citizenship of the child, the inadequate compensation paid to and lack of bargaining power of surrogates. The rights vs choice arguments do not really apply to infertile women, in India, both on the demand and on the supply side. The glorified motherhood and desperate concern over having heir to the family compel women to essentially make a choice, to have a child by compulsion and force and as such it is to be questioned as to whether it has to be termed as choice. Similarly, the 'rights' argument where it is not often asked whether the women have the right to say 'NO,' to infertility treatment or whether she is psychologically prepared or not, whether she wishes to have her own biological child or resort to adoption, whether she made completely 'informed choice' as the whole process, the kind of physical pain, number of attempts to be made to get the conception etc, are presented before a woman, going to the treatment. In the same line, the surrogate mothers' perception is to be understood as to whether she had the preference on her own to serve as a surrogate or forced, whether she expressed her right when the choice was made. The recent surrogate bill speaks about such rights of surrogate mother and recognizing the commercial aspects, now it is suggested that surrogate mothers to be hired from own blood relatives. Hence the choice and rights aspects to be dealt with taking the above factors and the policies must be framed accordingly, particularly taking into consideration the social pressures on women to procreate.

Shulamith Firestone (1971) feels that these technologies have the potential to liberate women, from the burden of motherhood and hence act as an instrument through which women's emancipation can be achieved. For supporters of this view, ARTs provide solutions for individual women, in a context, where infertility and childlessness result in social ridicule. Moreover, they empower women by giving them 'choice' and 'control' over their own reproduction. Such perception may be true in the West but in the Indian context, it is quite rare that any women get the child birth outsourced in the name of freedom of women. Though the technology does inform that such choices are available, in fact, there were cases in certain contexts, where it was attempted by some working women and in certain context, for not losing the beauty of women, surrogacy was sought for. But it was put to serious criticisms.

Sistare Christine (1988), has written that a "fundamental moral issue in the surrogacy debate is the nature and extent of women's freedom, their freedom to control their bodies, their lives, their reproductive powers, and to determine the social use of those reproductive capacities."

Influence for Surrogacy

There are different schools of thought, favoring or disfavoring surrogacy. The mainstream technologists' arguments are in favor of surrogacy and some of the feminists do argue that it frees women from bearing the child and the associated pain. But there are strong feminist critiques that surrogacy could reduce women to mere reproductive role of women and they could be pressurized into procreating, without opening or utilizing the choice, to remain childless or adopt a child. It is also favored in the context where the sex non-conforming group, could get a child, through ARTs.

Agigian (2004) opined that the assisted reproduction has brought increased freedom and opportunity for some people, making parenthood possible for infertile couples, single men and women, and gay and lesbian couples, through artificial insemination, surrogacy, or *IVF*.

Amrita Pande (2009), in her study, focused on surrogacy as labor. Most surrogates and their families do not recognize surrogacy, as paid labor, performed by women. The in laws and husbands of the surrogates perceive surrogacy as a familial obligation and a duty. The striking absence of surrogacy, as work, in the narratives of the surrogates indicates that the surrogates do not resist this image of women as selfless dutiful women, whose primary role is to serve the family. It is quite a differing and acceptable view point, where a woman hiring the uterus is just considered as paid work where the physical and mental pains are experienced like in any other job. There are debates about paid and unpaid work and the labour market perspective and economic perspective are that any work, which has monetary value, is productive work and which falls under the category of paid work. In such a context, sex work is also a work and also surrogacy where the work is being paid. Hence surrogacy cannot be overlooked in a labour market perspective as paid work.

Glenn (1992) held similar a view. Reproductive labor typically includes activities such as purchasing household goods, preparing and serving food, laundering and repairing clothing, socializing children and providing care and emotional support. Surrogacy can also be a part of the reproductive labour it implies.

According to **Radin (1996)**, surrogacy, especially in an international context, deconstructs the traditional nuclear family and offers alternatives that are potentially more liberating for women. Because of new technological advances, surrogacy also provides women direct access to contract, access that was severely limited in the past.

Spar (2006) stated that women were more willing to donate eggs, if they did not also have to undergo the pregnancy, and they were more interested in serving as surrogates if the child they were carrying, was not genetically theirs. Given the binding that if the children are genetically the woman's, it will have a different bonding. So surrogacy terms and conditions typically underline a contract, where there is a kind of detachment, between the baby borne for commissioning parents as surrogate from the genetically bearing. It is this gap between the two that is being marketed.

Gender Perspective

Gender perspective is building and understanding from the perspective of male and female, which would reflect gender neutrality. There are studies, conducted with the gender perspective and they are presented below.

Anderson (1993) considers surrogacy as the extreme form of alienated labour, which is more about generating profits and reproducing sexism, rather than about generating life. 'When market norms are applied to the ways we allocate and understand parental rights and responsibilities over children, children are reduced from subjects of love to objects of use. When market norms are applied to the ways we treat and understand women's reproductive labour, women are reduced from subjects of respect and consideration to objects of use.' This has been the arguments against surrogacy, where a human being is reduced to a product or a commodity and focus would be marketed as a product. Surrogacy dynamics and the process very well evidence such facts.

Crozier (2010) states that, payment for reproductive services is problematic because ova retrieval and pregnancy are physically invasive and involve significant health risks. Concerns are more serious, when women in financial need, resort to these practices for financial compensation, in which case there is no real autonomy in their decision-making. These are enhanced when practices go beyond national borders and into a global market, where consumers are wealthy people from developed countries and providers are poor women in the third world. Such a gap in economic status, between the third world and rich countries, does have an influence over the strong bargaining through agents and middlemen, for the commissioning parents, who after all come to India for getting the baby with a lower cost. At the same time, the poor are put up at the weak bargaining as they agreed to surrogacy for pressure of economic needs and forced to accept whatever is being offered. Such poor economic status is being exploited by the market and sometimes, where the surrogate women are given lump sum, which they would not have received in their life time. It is also equally true that the women's body has been involved in the transaction which the women themselves failed to recognize because it is rather a great contribution that uterus is being rented out, with all pains.

Daisy Deomampo (2013) analyzes ethnographic data, within a geographical framework, to examine how actors embody and experience power relations through space and movement, revealing how power is not simply about who moves and who doesn't. The author also discussed the reproductive practices and existing unequal relations, that reflect the contours of stratified reproduction. Also, reproductive tourism builds on feminist theoretical analyses of reproductive labor and social stratification, particularly within the landscape of global capitalism, in the Twenty First Century. Reproductive tourism has been seriously taken and in the new Surrogacy bill, it has been addressed from the surrogate perspective.

Elizabeth Anderson (1990) echoes this objection, but adds that surrogacy contracts also alienate a woman from her love for the child and frequently involve exploitation, as surrogate sellers have less wealth and are more emotionally vulnerable than buyers. Other objections stress the weakening of the link between parent and child, and the special vulnerability of children. It is complete detachment between mother and the baby, which is being thoroughly insisted throughout the surrogate process and hence the surrogate mothers do not develop any emotional bonding with the child they had borne. Market will kill the humanity and make a person who bear a child also, to remain extremely detached or do not allow them to express their emotions.

According to **Gupta (2006)**, the translational flow of reproductive products like gametes and embryos, re-emphasizes pre existing power imbalances, with globalization creating new regimes of consumption, driving the reposition of women in new systems of inequality. It is true that the women are just treated as object, in the consumption process, where the money is invested to consume the services of surrogate mothers

According to **Gupta and Richters (2008)** stated that ARTs can be understood as encompassing political, economic, ethical, cultural, and social processes, thus raising questions of gender, family, bioethics, law, and feminism. While these technologies originated in the Global North, their spread to the Global South was rapid, accelerated in more recent years by globalization and the neoliberal doctrine of trade without borders. Globalization is capable of promoting technology-based innovations and using the third world countries for market, to earn profit where the surrogate mothers are treated as objects, to deliver the child.

According to **John Stuart Mill's (1859),** only harmful practices should be prohibited by law and that one is ultimately sovereign over one's body and mind. The new Surrogacy Bill 2015 very much underlines such notions and does not promote commercial surrogacy but only through relatives.

Kimbrell (1993) draws parallels between surrogacy and slavery, since slave women were often used as birth mothers without any legal rights. Fears are expressed that poor women might be transformed into an army of surrogate labour or a caste of pregnancy carriers. As there is a slavery attitude being developed, it may not be a surprise when the surrogate mothers would be perceived as slaves.

Radin (1993) held that the feminists, who do want to fully legalize surrogacy follow the reasoning that the world is non-ideal. Women and men are not equal and for years, women have been relegated to a separate sphere at home, away from the marketplace. This has made women powerless because the place of power is the marketplace, which is dominated by men. This power has meant the liberation of men. Women want to achieve this as well. They do not want men to tell them what to sell and what not to. Whether or not it is morally wrong to engage in child selling and surrogacy, should be decided by the women themselves. But in the Indian context, it may not be possible and ideologically, women may not agree immediately to such a practice. If accepted for economic reasons, it has to be with the consent of the members of the family, her husband, where the surrogate mothers are to convince them. Women have no choice due to pressure of debts etc and they are willing to accept surrogacy, without being forced by others in India.

Rosemarie Tong (1990) says that surrogacy is harmful in straight forward non moral terms: "since there is evidence that surrogacy arrangements harm contracted mothers, a ban on commercial surrogacy needs to rely only on the harm principle" (as opposed to legal moralism).

Caste, Class and Gender Inter-Sectionalities

Ryan (2009) finds that feminists working in developing countries or economically impoverished regions, fight against a generalized "single-issue" approach, to reproductive policies and philosophies and call for policies regarding ART to better understand the intersection of "poverty, gender, inequality, and cultural religious norms, underlying sexual and reproductive health challenges." It is such intersections which primarily decide the surrogate mothers to serve as surrogates.

Sama (2010) found that the ART industry holds the promise of not just delivering babies, but also of enabling the reproduction of the desired caste, religion, class, etc. Therefore, it becomes imperative to examine the intersections between caste, gender, religion, class, region, and profession, to understand the complexity of ARTs. It is called designer baby where the commissioning parents do express their preferences towards sex, color, origin of the surrogate, religion etc.

Satz (1992) explained that there are limits to the objections based on an intimate connection between reproductive labor and our selves. Satz's argument stresses the background context of commercial surrogacy i.e, the gender inequality in modern society. Commercial surrogacy allows women's labour to be used and controlled by others, and reinforces stereotypes about women. He also explained that pregnancy contracts give buyers substantial control rights over women's bodies and rights to determine what the women eat, drink and do. They also may deepen stereotypes that women are baby-machines. Finally, the race and class dimensions of such markets also need to be considered.

According to **Scott (2009),** the superior educational and economic resources of the commissioning parents, virtually guarantee that the negotiation situation will favor them. Unless the surrogate herself is proactive in advocating her interests, no one in the negotiation has an incentive to ensure that the surrogates do not bear the full weight of the risks associated with these agreements.

Shore (1992) argues that not only do they crystallize issues at the heart of contemporary social and political struggles over sexuality, reproduction, gender relations and the family but also they challenge our most established ideas about motherhood, paternity, biological inheritance, the integrity of the family, and the 'naturalness' of birth itself.

To sum up, it is well revealed, in the studies presented as review of literature, that there are various dimensions of the surrogacy and the ARTs. All the studies have been categorized and presented with an analysis. The feminist perspective based studies, questioned the very issue of treating women's body as a commodity or a machine. From the technological perspectives, it has been perceived as a blessing for the childless couples. There has been supporters and the studies in such line underlined that the ARTs do help infertile couple and other single persons to have a child, while from the surrogate mothers' perspectives, it has been quite focusing on their rights and exploitation of their vulnerability, by various stakeholders involved right from the agents, brokers, commissioning parents, doctors, mediators etc. All said and done, in a balanced perspective it may be justified that the technology serves the purpose of infertility, which is a curse in the Indian context and it has lot of implications for the status of women. While in a feminist perspective, women are reduced to a machine or commodity, on the other side, women, in the patriarchal society, get recognized for their reproductive role and if that is being outsourced, the existing situation of subordination will be in question. The review of literature has served as a base, for proceeding further, to place the objectives of the study. Moreover, majority of the studies reviewed have been done in other countries and only a very few studies have been conducted in India and as such it has been quite relevant and appropriate for the present study.

Chapter IV

Operational Definitions of Technical Concepts and Feminist Theories for Surrogacy

This Chapter brings out the definitions of the technical concepts and the feminist theories. There are several technical concepts which are to be defined, for an understanding of the technical details of surrogacy. The feminist theories, which have influence over the decision for woman to serve as surrogate must also be discussed. Several technical concepts have been used in the study and they have defined, to help readers to understand the context in which specific concepts were used. This is explained in Section I, followed by the Section II, presenting the feminist theories. The theoretical framework of the study has been presented, with the application drawn from the feminist theories.

Section I: - Technical Concepts and Operational Definitions

The definition of the technical concepts, used in general and what they mean in this study, are presented below.

Alienation: It simply means isolated. Persons are isolated from society economically and emotionally and they do not have any platform to express their feeling like pleasure and pain, neglected from health care, family concern etc.

Biological Determinism: Biological Determinism refers to the phenomenon where certain biological features determine the totality of ones self and that physiological differences between men and women determine their social roles.

Capitalism: A form of economic order, characterized by private ownership of the means of production and the freedom of private owners to use, buy and sell their property or services in the market, at voluntarily agreed prices and terms, with only minimal interference with such transactions by the State or other authoritative third parties.

Control of Female Sexuality: The control of female sexuality is a crucial element in patriarchy. Female sexuality is regulated and controlled by male i.e. sexual desire, sexual preference, sexual pleasures, and sexual practice are decided and handled by male, for their own gratification and needs of female are not considered as subjects and treated as mere objects.

Commoditization: Commoditization means transfer of things mutually such as goods and service or others. It is defined as commoditization of women's body particularly reducing 'uterus' of women to a commodity and fix the price for renting the uterus as a product.

Control: Restricting oneself to act freely, by using physical, verbal, sexual and economic force. It is keeping others in control without power and freedom. In this study the surrogate mothers are completely held up as per the agreements right from conception till delivery without any freedom to act and almost imprisoned, stay away from normal life, family, children etc.

Culture: The full range of learned behavior patterns that are acquired by people, as members of a particular community. A culture is a complex, largely interconnected whole that consists of the knowledge, belief, art, law, morals, customs, skills, and habits, learned from parents and others, in the community. Culture is the primary adaptive mechanism for humans. Culture is defined by the society and social institutions, say religion, caste, etc. In this study, culture is defined in terms of the patriarchal perspective where the man decides and women just implement. Surrogacy is the other way where the market rules and motivates the women to take a decision and get it permitted by men, to hire out the uterus where the women make a choice by pressurized with economic needs to earn money and humanistic concerns to help the childless couple to have a child.

Decent Work: International Labor Organization (ILO) defines "decent work" as secure work that respects core labor standards, guarantees adequate income and social security and respects the freedom to enter into social dialogue as well as the freedom of association. Surrogacy is viewed as labour where the entire services of reproductive system are appointed to deliver the labour required and payment is made. But it cannot be brought under labour law, given the nature of transaction.

Devalue: Reduce or underestimate the worth or importance of. In this study it is the women's body is taken to a product and her services as just labour to conceive and deliver a baby without claiming any ownership or association post delivery.

Discrimination: The act of showing partiality or prejudice, a prejudicial act, unequal treatment, of men and women. In the Women's Studies context, it refers to the gender discrimination where there men and women are given to operate in different environment, focus, opportunity, recognition, pay etc. Discrimination of surrogate mothers as just a contract labourer is visualized in the surrogate market.

Domestic Violence: It is a pattern of abusive behavior, in any relationship, that is used by one partner to gain or maintain power and control over another intimate partner. Domestic violence can be physical, sexual, emotional, economic or psychological actions or threats of actions that influence another person. This includes any behavior that isolates, intimidates, manipulates, frightens, humiliates, terrorizes, coerces, and threatens to someone. Domestic violence, loss of earnings from male members do affect the economic needs and women are forced to opt for surrogacy. In such a volatile situation, it is quite difficult to be guaranteed that surrogacy service rendered by women might have taken to domestic violence.

Economic Abuse: It is defined as making or attempting to make an individual, financially dependent, by maintaining total control over financial resources or withholding one's access to money. Surrogate mothers are often willingly take up the assignment or contract for payment and hire out their services. But it is also forced labour in several context, either by agents, peers or family members.

Emotional Abuse: Undermining an individual's sense of self-worth and/or self-esteem. This may include, but is not limited to, constant criticism, diminishing one's abilities, name-calling, or damaging one's relationship with his or her children. The post delivery and away from family by the surrogates do affect their emotional abuse when the conception fails, or do not adhere to any form of stipulations by the market.

Empowerment: Empowerment is based on the idea that giving employees' skills, resources, authority, opportunity, motivation as well as holding them responsible and accountable for outcomes of their actions, will contribute to their competence and satisfaction. It is to provide them the ability to take decision on their own. In the Women's Studies context, it is sharing of power and providing an enabling environment to work to ensure gender mainstreaming with the focus on economic social and political upliftment of women to take decisions.

Endogamous Marriage: Endogamous marriage is intra marriage or inside marriage. Therefore, endogamy is the type of marriage which takes place within the same group, class, caste, religion, tribe etc, and both the partners belong to the same class or caste. Endogamous marriage is that which is controlled by the same group. Surrogacy just deviates from endogamous child birth as far as the child is being borne by other women, though it is endogamous for commissioning parents.

Exploitation: Misuse of one person or group by another. In economics, it refers to the appropriation of a person's or group's labor by someone else. Example: The capitalist exploits the worker by appropriating the product of his/her labor. In surrogacy, the question of exploitation comes in the form of surrogate mother being cheated with low payment or not making the payment in time or not making the agreed amount where she is put to exploitation

Feminine: Feminine is characteristic of female gender performance. Feminine is also often used to describe a lack of masculine performance from a male-bodied person. It is a myth and it is society created. There is always some feminine and masculine qualities present with every woman and man and no need to have any categorical divide. In infertility context, the men are perceived as not providing their masculinity if they are infertile, and women are excluded and alienated, ill treated when infertile.

Feminism is an awareness of patriarchal control, exploitation and oppression of material and ideological levels of women's labour, fertility and sexuality in the family, at the place of work, and in society in general and conscious action by women and men to transform the present situation.

Feminist Research Methodology (FRM): Feminist research is characterized by its double dimension. As opposed to traditional research, its objectives include both the construction of new knowledge and the production of social change. Also, feminist research is grounded in feminist values and beliefs. Feminist research is characterized by its diversity. It is interdisciplinary and transdisciplinary. It uses different methodologies and it is constantly being redefined by the concerns of women, coming from very different perspectives.

Feminization of Poverty: The feminization of poverty is a change in the levels of poverty, biased against women or female headed households. More specifically, it is an increase in the difference in the levels of poverty among female versus male and couple headed households. It can also mean an increase of the role that gender inequalities have become a determinant of poverty, which would characterize a feminization of the causes of poverty.

Gender Analysis: Gender Analysis refers to the variety of methods, used to understand the relationships, between men and women, their access to resources, their activities and the constraints they face relative to each other. Analyzing, both the male and female, with gendered lens.

Gender Discrimination: Gender Discrimination or sexism is prejudice or discrimination based on a person's sex or gender. Sexism can affect any gender but it is particularly documented as affecting women and girls.

Gender-Based Violence (GBV): Gender-based violence is a human rights violation, a public health challenge, and a barrier to civic, social, political, and economic participation.

Institutional Oppression: Arrangement of a society, used to benefit one group, at the expense of another, through the use of language, media education, religion, economics, etc. Surrogacy had been reported right from epic era where the serving community women were expected to deliver the child in order to breast feed the children of the queen. In the absence of a male heir, there were informal arrangements whereby the was in relationship with other women than wives, to get a baby. This institutionalized practice and oppression of women, after the delivery for breast feeding, happened often.

Invisible Minority: A group whose minority status is not always immediately visible, such as some disabled people and LGBTIQ group. This lack of visibility may make organizing for their rights difficult. It may include religious, racial, ethnic minorities who would face all related merits and demerits of being a minority.

Male Supremacy: A system of oppression, that gives power to men and privileges masculinity, at the expense of women and femininity.

Marginalization: The process whereby something or someone is pushed to edge of a group and accorded lesser importance. This is predominantly a social phenomenon by which a minority or sub group is excluded and their needs or desires ignored.

Marginalized: Excluded, ignored or relegated to the outer edge of a group/society/community. Alienated from the centralized focus of the community.

Masculine: Masculine is characteristic of or befitting male gender performance. Masculine is also often used to describe a lack of feminine performance in a female-bodied person.

Migration: Migration refers to the movement of people from one place to another for fulfilling their needs.

Oppression: Results from the use of institutional power and privilege where one person or group benefits at the expense of another. Oppression is the use of power and the effects of domination.

Patriarchy: It refers to male dominated society. It is the systemization of the oppression of women by social structures such as marriage, heterosexuality, law, politics, and even language.

Personal is Political: This phrase refers to the experiences, feelings, and possibilities of our personal lives, which were not just a matter of personal preferences and choices but were limited, molded, and defined by the broader political and social setting. They feel personal, and their details are personal, but their broad texture and character, and especially the limits within which these evolve, are largely systemic.

Psychological Abuse: Elements of psychological abuse include, but are not limited to, causing fear by intimidation, threatening physical harm to self, partner, children, or partner's family or friends, destruction of pets and property, and forcing isolation from family, friends, or school or work

Sexism: Discrimination against women that results from the male supremacy system of oppression. It is expected that only women must bear the brunt of infertility and hence all sort of methods to get a child is being followed.

Slavery: The subjection of a person to another person, especially in being forced into work and work done in harsh conditions for low pay. The dominant group acts as the master of the cradle. Here domination means the economic supremacy of people. Surrogate mothers are enslaved to agents and middlemen, who bring them to the contract and they are at the mercy of such agents, in getting their payment made.

Social Cost: It refers to the total expenditure of the society such as action, event and policy change, where the damage to the society or any cost involved to create facility where if it does not contribute, will incur a social cost. Surrogacy is a social cost for infertility where the women for the sake of addressing the infertility issues, hire out her uterus. It is painful but still for the social cause, it is being done by surrogate mothers

Subjugation: To make submissive or subservient. Surrogate mother, egg donors etc are put into vulnerability and subjugation without any freedom. It is medically insisted and also required for successful completion of the surrogacy process.

Systems of Oppression: Interlocking societal, economic, moral, and religious values that keep many groups of people down, to ensure the power and advantage of a few groups or one group of people. Some systems of oppression in the US and Europe include heterosexism, male supremacy, white supremacy and the gender binary system. Patriarchy insists on reproductive role of women and does not allow an infertile woman to resort to adoption but to go for biological child and hence surrogate mothers agree to serve. Women mostly undergo systemic oppression which is also faced by lower caste, class, minority, rural, less educated, disabled and other marginalized.

Transgender: An umbrella term for transsexuals, cross-dressers (transvestites), transgenderists, gender queers, and people who identify themselves as neither female nor male. Transgender is not a sexual orientation. Transgender people may have any sexual orientation. It is important to acknowledge that while some people may fit under this definition of transgender, they may not identify as such. There is facility for transgender or other LGBQT Group, to have a child, with surrogacy arrangement.

Section II

Altruistic Surrogacy: An altruistic surrogate is one who selflessly volunteers her reproductive services to another, without receiving any sort of monetary gain in return. Under this surrogacy, the surrogate might be family members, relatives and friends (within blood relations).

Artificial Insemination (AI): Artificial Insemination is the procedure of transferring semen into the reproductive system of a woman. This technique comprises of artificial insemination with husband's (AIH) or with donor sperm (AID)

Assisted Reproductive Technology (ARTs): Assisted Reproductive Technologies assist in conception or in the carrying of pregnancy. If unchecked, it could become a veritable 'fertility industry.'

Blastocyst Transfer: Embryos are grown to a more advanced, stronger stage, very similar to the stage when embryos are implanted in the uterus, and are transferred five days after the egg retrieval.

Caesarean Section: A surgical incision of the walls of the abdomen and uterus for delivery of offspring.

Commercial Surrogacy: It is similar to a business agreement and includes a contract between the surrogate and the intended parents, which includes coverage of her medical coast and payment for her services. Although both share the common goal of providing reproductively challenged people with children of their own, the two are perceived quite differently. The compensation that the surrogate is provided through commercial surrogacy, leads to its association with the commoditization of reproduction and the notion of buying and selling as forms of property. (Edelmann 2004:152).

Commissioning Couples: Commissioning Couples are those who wish to have their own genetic child.

Commercial Surrogacy: In commercial surrogacy, the surrogate's eggs are not used at all. Therefore, the child will not be related to the surrogate biologically. Using the commercial type of surrogacy, the embryo is actually created by using both the biological father's sperm and the biological mother's egg through a process called *in vitro* fertilization.

Cryopreservation: A method of preserving tissues (including sperm and eggs) and organs in a viable state, by freezing, at extremely low temperatures.

Embryo Transfer/Implantation/Transplant: The transfer of an embryo from *in vitro*culture into the uterus.

Embryo: The fertilized ovum that has begun cellular division and continued development, up to the blastocyst stage till the end of eight weeks.

Fertility: Fertility is the ability to have babies or to reproduce.

Gross Domestic Product (GDP): The total market value of all final goods and services, produced in a country in a given year, equal to total consumer, investment and government spending, plus the value of exports and minus the value of imports

Heterosexual: Heterosexual is sexual attraction or sexual behavior between persons of opposite sex or gender.

Homosexual: Homosexual is sexually attracted to people of the same sex or gender

Infertility: Infertility is a disease of the reproductive system, defined by the failure to achieve a clinical pregnancy, after 12 months or more of regular, unprotected, sexual intercourse (WHO-ICMAR Glossary)

Intermediates/Agents/Brokers: Theyfunction as intermediaries, between two or more parties, in negotiating agreements, bargains, or the like.

Intra Cytoplasmic Sperm Injection (ICSI): Injection by a micro-needle of a single sperm into an egg, followed by transfer of the egg to an incubator where fertilization takes place and then introduction of the fertilized egg into a female's uterus. Used most commonly in cases of male infertility or where the oocytes cannot be easily penetrated by sperm.

Intra Uterine Insemination (IUI): Placement of washed sperm into the uterus.

IVF-ET (*In Vitro* Fertilization - Embryo Transfer): *In-Vitro* Fertilization-Embryo Transfer is the fertilization of an ovum outside the body and the transfer of the fertilized embryo to the uterus of a woman.

Kinship/Family: The relationship between members of the family (Blood Relations).

Miscarriage: A miscarriage is the loss of a fetus, from natural causes, before the twentieth week of pregnancy.

Motherhood: Motherhood is the state or experience of having and raising a child.

Multiple Pregnancies/Multiple Gestation: The condition of having more than one fetus in the uterus.

Oocyte Retrieval (Egg Retrieval): Process of removal of the egg, by the technique of aspiration from the ovaries.

Oocyte: The female sex cell (ovum) produced by the ovary, which when fertilized, produces an embryo.

Ovarian Cyst: A benign or malignant growth on an ovary.

Pregnancy: The state of carrying and developing embryo or fetus within the female body

Prolactin: A protein hormone, released by the pituitary gland, that induces and regulates lactation.

Reproduction: Reproduction is any form of sexual reproduction, resulting in human fertilization, typically involving sexual intercourse between a man and a woman. During sexual intercourse, the interaction between the male and female reproductive systems, results in fertilization of the woman's ovum by the man's sperm.

Reproductive Health: Reproductive Health implies that people are able to have a responsible, satisfying and safe sex life and that they have the capability to reproduce and the freedom to decide if, when and how often to do so (WHO).

Reproductive Tract Infection (RTI): Infections that affect the reproductive tract. In women, reproductive tract infections can be either in the upper reproductive tract (fallopian tubes, ovary and uterus) or, and the lower reproductive tract (vagina, cervix and vulva) and in men, these infections are at the penis, testicles, urethra or the sperm tube.

Sperm: The male reproductive cell that fertilizes a woman's egg. The sperm head carries genetic material (chromosomes), the mid-piece produces energy for movement, and the long, thin tail propels the sperm.

Still Birth: A still birth is the loss of a fetus, from natural causes, after 20th week pregnancy.

Surrogacy: Surrogacy is a process by which a woman becomes pregnant and gives birth to a baby, in order to give it to someone, who cannot have children on their own.

Surrogate Mother: Women who bear a child, on behalf of another woman, is called Surrogate Mother.

The concepts related to social, cultural, gender, reproductive, medical terminologies have been defined thoroughly, to fill the gap and make the readers to have proper understanding on the concepts. Surrogacy is governed by social, political, medical and technical aspects and hence a thorough definition has been given.

Section III: - Theoretical Framework

Surrogacy is helping the infertile couples to have their biological child. Commissioning parents long for a child especially their own biological child. Hence they seek surrogacy as their source, to get over infertility. The demand from such mothers is well capitalized and converted into a marketable product by ARTs and medical tourism industry has come to flourish. The ARTs have been brought to the market, by the Gynecologists or the Medical industry, who prefer the surrogacy. In the surrogate market, the price is fixed by the agents, intermediaries, exploiting both the commissioning parents and the surrogate mothers. The former is fleeced by the agents, by using their anxiety to have a child of their own and the surrogate mothers are exploited due to their economic vulnerability. With the given Indian cultural principle, no woman is encouraged to donate or take the sperm from other men and it is labeled as prostitution. But technology breaks all such beliefs and the market really helps to come out of such beliefs and offers scope for both the demanding commissioning parents and the supply side of surrogate mothers at a point where both their needs are satisfied. Market does not have any sentiments nor any human considerations or ethics and hence a detachment happens once the child is born. The market makes no arrangement for post natal care, psychological needs, trauma etc for the surrogate mothers and throws them out by just paying a lump sum, received from the commissioning parents. The trio of commissioning parents the market and the surrogates form a clear equation, to capitalize the market for infertility treatment. Feminist concerns about surrogacy reducing women to a state of a machine, reinforcing the stereotyped role of women as glorified motherhood etc, were taken as the research questions. The intersection of the market and technology, with the fertility, has been analyzed, from women's rights perspectives and humanistic concern.

There is no doubt that the question of infertility and technological advancement, in the reproductive health, must be discussed as a global issue. The technology must be used for the development of human kind, but it is to be questioned as to whether every technology is introduced only for construction and not destruction. Any technology, which is to serve humanity, should not become commercial and exploit the end users. The ARTs, particularly surrogacy, must be analyzed within such a framework, as to how the social pressure on childless couples exposes to social taboos, for not being able to have a child. The patriarchy is winning in every sphere, putting women in a state of subordination and disadvantage and facilitating men to be in the comfort and safe zone. The ARTs and its impact need not be analyzed in the context where the couples are ready for adoption, but the very existence of such ARTs make the option for adoption a secondary one.

Feminist Theories for Surrogacy

Theoretical background is more important to understand the issues in the society. Feminist theories are liberal and radical and both these theories have been equally discussed, over the surrogacy market. The radical feminist theories appreciate the application of ARTs as it relieves the women from bearing the child and surrogacy may help them to get their biological child. LGBQT and single woman can get a child. However, it is equally condemned because women are reduced to the reproduction alone, though women's agency could contribute and achieve, in all possible areas of domains, which are still male centric and occupied by men. The liberal feminist school emphasizes that women can participate in public life after fulfilling their reproductive role. The following section discusses the feminist theories, which will provide the base, for surrogate markets.

Marxist Theory of Exploitation

Marxist theory focuses on the principle of exploitation. Exploitation is defined as one's vulnerability, which is used by others, for their benefits. Benefitting or getting help from one another, does not come under exploitation but to some extent, the concept was totally changed in surrogacy. One's weakness should not be used by others for their survival, in the name of commoditization. Karl Marx believed that economic and political institutions of capitalism were exploitative in nature. Exploitation can also be harmless or mutually beneficial. If the exploitation had been harmless or mutually beneficial, there is no issue. But if the same victimizes the individual or violates the rights of the individual, we need to look at the issue seriously. Surrogacy is viewed as mutually supporting the commissioning parents and surrogate mothers. As long as it is mutually supporting, it is not to be viewed as exploitation. But in the name of mediators, agents, the actual amount to be paid as agreed, or actually paid by the commissioning parents, if not paid or the actual amount to be paid by the commissioning parents is charged more than what is actually agreed to be paid to the doctors, surrogate mothers, it amounts to exploitation. Different kinds of exploitation are discussed by Marxists, such as exploitation with harm, without harm, sexual division of labor, bonded labor, capitalism etc., Marx strongly insisted that in a capitalistic society, the exploited are forced to sell their labor power to capitalists, for less than the full value of the commodities they produced, with their labor. It is applicable to the surrogacy context also. As far as the commercial surrogacy is concerned, it constitutes multiple exploitation in the society. The reproductive health of women is commodified by the capitalistic society. Reproductive health of women is seen as a priceless thing. But the revolution of reproductive technology destroys the moral values in the society, under the capitalist society and attaches a price tag to the reproductive process.

The capitalist society creates multiple exploitations among women, in the name of commercial surrogacy, in several dimensions like economic, health and caste. Surrogacy is a process by which child is carried for a third party, for some remuneration. Carrying another's child is not an easy task but at the end of the contract, women are cheated by intermediates as well as by biological parents. After the successful implantation of embryos, the contract is made between the biological parents and surrogate mother, without their meeting in person, with the help of intermediates like brokers and agents. After the implantation, the contract is made for some remuneration, with the help of mediators. In some cases, an advance amount was claimed by the surrogates, to tide over their economical crisis. Very few of them collected remuneration completely, at the end of the process. Surrogates often collected twice, thrice or more than that and as a result, money was used like a pocket money by them. Majority of surrogates were involved in surrogacy because of attraction of huge money to clear their debt, but the installments did not completely fulfill their economical needs at the end of surrogacy. Surrogates were cheated by the intermediates and also to some extent, they were cheated by the biological parents. The intermediates collected one third of the amount from the biological parents, without the knowledge of surrogates, for their mediation process. Because of that, surrogates could not get their remuneration from the biological parents. On the other hand, at the end of surrogacy, biological parents did not pay the amount completely because the biological parents did not have money and they promised to pay later.

As Marx asserts, exploitation is not just in money terms or in working hours etc, it is also on race, caste, religion etc. It is true that in surrogacy, racist exploitation does happen. It occurs in the overseas. Further, caste based exploitation occurs in India, which is observed in surrogacy. There are different prices paid to different castes. The higher caste women are paid more than the lower caste women. But in general, it is the dalit women who are engaged in more number surrogacy contract. It was informed by the doctors that the commissioning parents do demand male babies, fair babies, babies from higher caste hierarchy etc. It is called "designer baby." During surrogacy process, based on the social caste, the remuneration was fixed by the intermediaries. The so-called upper caste women received more remuneration compared to so-called lower caste women. Also, due to vulnerability and feminization of poverty, women entered into discriminatory practices. In addition to caste, education and skin tone also played a major role in the fixing of remuneration for the surrogacy process.

Surrogacy creates health-related exploitation also. In surrogacy, healthy women alone were allowed to become a surrogate. For the benefit of biological parents, during the surrogacy, the women were forced to induce some biological activities in their body. These forced biological activities, pushed women into weaker position, after the surrogacy process. It is reported that the forced hormonal activities pushed women into vulnerable situation in their health, after the surrogacy. Capitalistic society creates labour in the name of employment. Feminization of poverty and vulnerable situation are used by the capitalistic society. Due to lack of power and capital, surrogacy creates sexual division of labour and bonded labour among women. The market economy attracted women to enter into bonded labour. The total working hours is eight in an industry but in bonded labour system, women are working in surrogacy, for twenty four hours. In surrogacy, bonded labour is leading to sexual division of labour. As discussed earlier, surrogate was not only forced into a contract due to financial insecurity but also, in some cases, subject to constant contractual restrictions, on her diet and activity. The sexual division of labour creates inequality and women are excluded in the society. Men also offer sperm donation but it does not create any harm nor did it lead to exploitation.

Hence it is clear that the surrogacy creates three different kinds of exploitation in economic, health and caste dimensions. The economic exploitation led to sexual division of labour and bonded labour. The labour system created slavery among women in the society. Also the caste wise exploitation les to alienation and the alienation caused social exclusion in the society. The alienated women were socially excluded from the society due to the moral values. Before and after surrogacy, health exploitation led to long term impairment of physical and mental health of the surrogate. Many surrogates also did not follow up physical and mental health care, after the surrogacy.

Patriarchal Theory on Surrogacy

The society is patriarchal and male centric. The sexual division of labour and the division of space into private and public, had not only vested enormous power on men, but also equally subordinated women. The combination of patriarchy – the family patriarch at the household front and capitalistic patriarch at work places, jointly works together to maintain the gender relations in the society, where the power differences continue to exist and expose women to exploitation.

Patriarchy is defined as the process of dominating and discriminating women by men in such a way that women are always subordinate in the patriarchal society, for the male benefits. There are different kinds of curse for women in the society under patriarchy, in the name of culture, religion, caste and ethics etc. Against this patriarchal background, after the marriage, women are forced to procreate and the woman, who is not able to reproduce due to some biological disorders, is socially excluded from the society. Feminist and well wishers of women argued that the reproductive capacity of women is the major reason for women to be subordinate to man in the society. Even women have internalized this value that women who cannot reproduce, are deemed "not perfect women" and she is not considered fit for life. Due to this reason, infertility is seen as the main and major curse for being a woman.

Infertility is seen as a major curse, for women in the patriarchal society. Infertility creates discrimination, subordination, oppression, and suppression among women. Before the globalization, infertility was seen as a major curse in women's life but after the globalization, liberation and privatization, the revolution of reproductive technology provided relief to women from the infertility curse. Procreative technology is seen as a social pain reliever and it makes entire life meaningful, to the vulnerable section under the patriarchal society.

It's interesting to note that technology does support patriarchy also and very often, it helps the male dominance to sustain. Technological revolution also creates discrimination against women. The technological reproduction stereotyped women's work. Procreation should be a choice for women but the patriarchal values insist that women should reproduce. Here, one question may arise about producing the child, for whom and to whom? The patriarchal society forced women, by cultural norms to accept honorable job for women. The women, who failed to procreate, were alienated and excluded from the society. Because of this reason, the reproductive technology is also called the Patriarchal Technology. The patriarchal technology questions the decision making power and especially women's reproductive rights. The woman, who is affected by infertility, is forced to use reproductive technology such as surrogacy. Soon, reproductive techniques offered surrogacy in commercial terms.

Surrogacy offers individuals an alternative form of family creation but it is riddled with issues of inequality and social injustice, based on gender, class, age, sexuality and geographical location. These social injustice issues are evident in surrogacy discourses, which have influenced public perception and particularly in relation to women and their role as mothers. Surrogacy is a complex issue as it contains various questions of power, body ownership, gender norms, parenthood, and global influences. In commercial surrogacy, women are considered to be under contract till they procreate. The commercial contract creates multiple questions among feminists such as, contract pregnancy gives others increased access to and control over women's bodies and sexuality. Secondly, commercial surrogacy reinforces the notion that women serve men as "baby machines." And third, it reinforces an old stereotype of women as merely the incubators of men's seed.

In addition to male domination and subordinate position of women, coupled with extensive violence and assault, there remains the larger possibility of emotional, familial pressure or coercion on the woman to act as a surrogate. Practicing of surrogacy creates questions of equity and equality between women in the society. If all women are equal in the society, then why women should be indirectly forced to undergo surrogacy by the capitalistic society.

Feminization of poverty means the burden of poverty is borne by women in the private and public life. In the developing countries, especially in India, there are so many reasons for women to be the victim of deprived economic situation. Firstly, there is economic dependency of women on the family. Secondly, patriarchic values deny women from social activities, particularly economic activity. Thirdly, women are alienated from a decent job because of lack of education, lack of required skills, lack of mobility and lack of awareness, family responsibility, social restriction and taboos. Fourth, women are considered as a reserve army and they are used when there is need for labour force, with minimum wage. Finally, in the public life, women are supposed to be victims of various forms of abuse and violence. Poverty happens in the families due to gender inequalities. There is difference between male and female headed households and based on the nature of households, the poverty levels also varied. In some situations economic dependency as well as the alcohol-addicted life partner, make the situation worse. In such a scenario, women are compelled to get money, from money lenders with higher interest, for their survival and the welfare of the family and children. This situation leads to economic crisis among women and to overcome the economical crisis, women are forced to undergo surrogacy.

Technology should be gender neutral but the reproductive technology is not gender neutral. It created discrimination between men and women and also between women. The reproductive technology, especially in the case of commercial surrogacy has oppressed, suppressed and marginalized women because of capitalistic structure. The most important thing is that patriarchal reproductive technologies create slaves among women. On the one hand, the reproductive technologies are produced by patriarchal society and indirectly forced women to access the technologies for their own benefits. On the other hand, religious leaders prevented women from accessing reproductive technologies because practicing those techniques are considered un-ethical. In such a scenario, producing, prohibiting and benefiting from procreative technologies, the patriarchal society also subjects women to abuse and exploitation. Women from lower caste or minority are paid low, treated with discrimination by the local intermediaries. In fact, such a background of women's social and economic vulnerability, has been well capitalized for the profitability of the agents or intermediaries, engaged in medical tourism.

To sum up, there are different forms of exploitation, which could very well be explained in the Marxist Feminist theoretical context, where it has been thoroughly discussed as how various forms of exploitation, apart from economic base, do exist which are highly gendered, male supportive and androcentric. Given that patriarchy is a capitalistic society has been a deciding factor in gender division of labour and decision making power, favoring the benefits of men alone, it is not a surprise that the surrogate market exploits women in various ways and makes them remain vulnerable and exploited forever.

Chapter V

Feminist Research Methodology and Research Design

"Not everything that can be counted,
counts, and not everything that counts, can be counted"
- (Albert Einstein)

The Researcher has adopted a narrative approach, to explain the experience of commercial surrogacy. The study has started with an aim to bring out the experiences of the surrogate mothers, from a feminist perspective, adopting feminist stand point theory. The feminist stand point approach, if applied to the surrogacy, would focus on the questions as to why women are still perceived only in procreation role.

They are human beings and given the changing social context, infertility need not be taken as an issue, where alternative means like adoption can be resorted to. Technological advancement should not reinforce the traditional gender roles. Women's body is targeted and treated as a product which cannot be entertained. The liberal schools of feminist held that women are reduced to mere reproductive role and fertility has become a huge public health issues. At the same time if at all the women do some get respect and recognition, it is the reproductive role, with procreation. With the emergence of ARTs and surrogacy, women may lose that recognition and may still be subordination. The radical feminists feel that the ARTs will relieve women from performing or pressurized to perform the procreative role and it facilitates the gender non-conformity persons, single women etc, to get an opportunity to get a child. Hence there are different views about surrogacy which must be discussed and analyzed from a feminist perspective.

The study has applied the Feminist Research Methodology in Women's Studies. Feminist research methodology is defined as research methods, which analyze and investigate from the feminist stand point and perspectives, by using feminist lens. It adopts qualitative techniques as well as focuses on an individual's experience.

Surrogate mothers and surrogacy are not mainstream process in our country. There are some social taboos, associated with the concept. For the purpose of identifying the participants, the Researcher used snow ball sampling technique, through which links and contacts were collected from one surrogate to another, surrogates who became agents and associations. For the purpose of research, the Researcher adopted non- probability method such as Snow Ball Sampling Technique, with the knowledge that it was not representative of the general population (Wimmer and Dominick, 2003). Qualitative research is an intensely personal and subjective style of research. The narrative approach is within the qualitative knowledge paradigm and centers on temporal stories in human experience. Surrogate motherhood is a temporal human experience wherein narrative is the way meaning is constructed (Polkinghorne, 1988). Narrative research does not search for universal truths but rather it looks for specific connections between events (Polkinghorne, 1988). In addition, narrative inquiry highlights ethical matters as well as shapes new theoretical understandings of people's experience (Clandinin & Huber, 2010, p. 436). Methodologically, the Researcher chose the narrative approach as it fitted with the study of meaning making in commercial surrogacy experiences. According to Polkinghorne, (2000), personal stories of lived experiences result in significant, yet often neglected, knowledge about human life. As such, personal and lived stories are events in which humans construct meaning through experience (Polkinghorne, 1988). Meaning making is an active human process, that occurs in the mental realm and requires linguistic data, to make relational operations and experiential implications explicit (Polkinghorne, 1988).

The topic chosen is to analyze the challenges of the surrogate mothers and the issues related to their rights. It was not an easy task to approach and get the data from the surrogate mothers as many did not want to reveal their identity. It was a herculean task to get appointment with one person and complete the data collection. Moreover, though the surrogate mothers were drawn from various regions in Tamil Nadu, majority of them performed their service at Chennai where they easily get contact and get motivated to accept the offer. Majority of women from rural areas got into surrogate motherhood with persuasion and they did not choose to reveal the fact to friends and relatives, except their own family members. Hence many surrogate mothers refused to provide data and those who accepted also, did not offer complete data. As an individual Researcher, it was not possible to convince them as there was no incentive, nor any need for them, to disclose their identities and co-operate with providing data. The Researcher really remains forever grateful to the Associations (G-Smart, Start India & Indian Surrogacy Law Center) who were kind enough to extend support and refer to nearly 75 surrogate mothers for interviewing and collecting data. The Researcher conducted interviews and recorded their interviews with the informed consent and kept the proceedings secret. However, suddenly there were issues with some of the members who were not paid the promised amount and there were other macro issues concerning surrogate mothers' rights and hence all the women had to proceed to participate in a protest programme, collectively organized. That was the end and it was never possible to get hold of the surrogate mothers again. The association facilitated that interview, which could not help as the Association was also under attack as the surrogate mothers were not happy and dissatisfied with the role of such associations in safeguarding their rights. Thus the interview phase came to a close to and the Researcher ended up with just 33 surrogate mothers with proper interview though it was ambitiously planned to interview more than 75 sample women. This was a serious a limitation of the exogenous study. However, care has been taken to present the analysis as intensive as possible, to reflect the views of other excluded women in the sample. Creswell (1988) suggested the range of 20-30 for grounded theory studies while Morse has suggested that within the range of 30-50, will be fine for a research where it is quite challenging to collect data.

Before presenting the objectives, methodology, it is worth presenting the research process in the studies of a qualitative and challenging kind. Hence, the Researcher represents the nature of care, rapport building, confidentiality to be maintained, informed consent to be taken, etc, in the following section.

Research Process

In order to conduct the study and collect data, the Researcher approached professionals in the fertility field, through particular ART clinics' mail ID and in person. Moreover, the Researcher contacted three Associations, which supported and counseled participants, in the capital of Tamil Nadu and requested their assistance for research. The Researcher discussed the research briefly with participants in person and by phone. Once the Researcher obtained permission from participants, the Researcher explained to each of them the aim and expectations of research.

Relationship - Building

Relationship-building was the key to this study design since participants were being asked highly personal and emotion - laden questions. Additionally, relationship-building was a part of a narrative approach in qualitative research. Bochner (cited in Newbury & Hoskins, 2010) suggests a relational approach to qualitative research, for generating new ways of comprehending subjective experiences, without documenting the results.

Ethical Considerations

Ethical considerations in research are critical. Ethics are the norms or standards for conduct, that distinguish between right and wrong. They help to determine the difference between acceptable and unacceptable behaviours.

Voluntary informed consent

The Researcher obtained verbal consent, at the beginning of every interview and received consent from all participants for audio taping, providing names or pseudonyms in the documentation of responses.

Pseudonyms

Due to ethical concern and to maintain confidentiality, the Researcher used pseudonyms. Additionally, all the names of family members, professionals, and children were eliminated or replaced by their designation.

Confidentiality

Confidentiality is another central ethical consideration for this study. Shultz Sheppard and Sheppard (2006) have maintained that confidentiality provides assurance of trust or confidence in the person with whom private matters are shared. In the spirit of maintaining confidentiality, the Researcher transcribed audio recordings herself and eliminated identifying information in narratives, as requested by participants. However, due to the small network of people involved in the world of commercial surrogacy, the Researcher informed participants that although confidentiality will be of the upmost importance in this research, it might not necessarily be maintained.

Reflexivity

According to Amulya (2006), reflexivity is shaped by questions, dialogue and narratives. Stories and dialogue can be effective technologies, for the reflective process because they provide cognitively complex and culturally potent systems, for conveying the way we think about, feel about and make connections to experience (Amulya, 2006). While writing the individual stories of the participants and using reflexive practice, emergent group themes and discourses became noteworthy knowledge. As White (2007) suggests, by critically reflecting on our intellectual traditions and ways of knowing, we are practicing an important form of accountability. The willingness of the Researcher, to use reflexivity, during the interviewing and interpretation processes, also contributed to credibility in this research. Reflexivity is thinking, doing, and acting critically (White, 2007). While questioning, interpreting, and analyzing participants' experiences and dialogue, the Researcher willingly practiced reflexivity, which established accountability and trustworthiness.

Establishing Trustworthiness

The Researcher attempted to stay as close to the data as possible, by transcribing the interviews *verbatim*. At the end of each interview, the Researcher made written notes about immediate observations and reactions from the interview. The rationale for commenting on each interview, was based on the understanding that analysis occurred throughout the research process and not only when all the data were collected. Weiss (1994) adds that researchers undoubtedly develop insights, theories, and the beginnings of analysis, during the first interview. Hence by writing initial reactions and observations after each interview, the Researcher began the analysis phase as well. According to Chamberlain (2000), while evaluating qualitative research, trustworthiness is necessary. While establishing further trustworthiness, the Researcher explained the context of this research and position in relation to the research topic, as Winter and Daniluk (2004) did in their study of egg donors.

The Researcher checked out some interpretations with some participants, the literature, own experience and dialogue with association members. According to Polkinghorne (1988), validity in narrative research is addressed by an inter-subjective consensus that minimizes falsity and arbitrariness in interpretation. This inter-subjective consensus is apparent in the telling of the surrogates' stories as they unknowingly follow a similar sequential pattern of their individual commercial surrogacy experiences, beginning with their first thoughts of surrogacy and the screening process while ending with the birth of the surrogate baby and their understanding of this experience. The individual stories mainly follow an analogous timeline, that includes similar themes and plots, thus establishing inter subjective consensus or coherence, as attested by Reissman (1993), in the data collection and interpretation. With similar thought, Averill, as cited by Churchill (2000), suggests that validity is always relative to the inferences being made.

As mentioned earlier, many of the surrogates indicated that it was important for them, in telling their personal stories, to position commercial surrogacy in a positive light. This created a challenge for the Researcher as a narrative researcher, who sought to analyze the stories from social, economic, religious, cultural and legal perspectives. Ethically, the Researcher wanted to respect the participants' wishes in representing their personal stories. However, as a Researcher, it was necessary to analyze the narratives without prejudice. In finding a balance between the tension of participant desires and analytic research, the Researcher elected to bring own narrative forward for the analysis. Researcher's interpretations for analysis, most likely with research perspective, may not align with theirs. With this background of information about the methodology, the following section presents the objectives, followed by hypotheses and methodology.

Objectives

- ✓ To study the socio economic, demographic characteristics of surrogate mothers.
- ✓ To identify the motivation, obstacles and facilitative sources to offer to surrogacy.
- ✓ To analyze the process of identification, persuasion and compromises the women made and the nature of market through which for surrogacy

emerged, and the locate the associated issues, if any, between the terms and conditions agreed upon and actually implemented, from a legal perspective.

- ✓ To find out the issues faced by the surrogate mothers post surrogacy – emotional, psychological, physical health and trauma and strategies to manage the same.
- ✓ To present the perception of various stakeholders like Doctors, Associations, Surrogate Mothers, Community and present the reflections on the New Draft ART Bill 2016, with the pre and post Bill changes and safeguarding mechanisms.
- ✓ To provide suggestions and recommendation based on the findings, to policy makers.

Methodology

The study was done in Tamil Nadu, as the sample State, where there was relatively higher expansion of reproductive health system, is taking place both in State and private institutions. The number of infertility treatment clinics has also been mushrooming where the Gynecologists do continuous research on application of various methods to treat and capitalize the growing menace of infertility. Having understood through the discussion and interaction with Associations, working for the rights of surrogate mothers and Doctors etc, that the surrogate mothers, though having from different parts of the State, they were primarily present in Chennai. Hence Chennai was identified as the geographical area to meet the surrogate mothers.

There were different attempts made, to trace the list of surrogate mothers. The infertility treatment clinics, across the State, were visited and discussions were held and requested for the list of surrogate mothers, but all went in vain as the doctors did not provide the list due to extreme confidentiality to be maintained. The Researcher promised that it was only for research purpose and data would be kept confidential but the Researcher did not succeed. One of the leading gynecologists, known to entire State, referred to a few surrogate mothers, who were also serving as agents to locate surrogate mothers in Salem and Nammakkal areas. Having made the contact, the Researcher approached the surrogate mothers (Agent) with appointment, and reached the place but information was not offered. The surrogate mother, who gave appointment to meet in a particular place, etc refrained from supporting the Researcher at the last minute. Other attempts to reach her were not successful. Hence it was given up. If that particular surrogate, who also served as agent, had been accessed, it would have provided an opportunity to meet the rural based surrogate mothers in places, other than Chennai, but it was not possible due to the non-co-operation of the participant. Hence the Researcher struggled, with all possible issues that researches face with respect to conducting social science research. Hence the Association, functioning for protecting the rights of surrogate mothers, was approached. The Association supported the Researcher by furnishing the list of surrogate mothers, who were the members of the Association. But, as mentioned already, interview process had to be stopped in the middle due to some issues among the members and the Researcher was forced to limit the sample to confine to the 33 surrogate mothers..

The socio economic analysis of the participants, the factors which motivated or forced them to accept to serve as surrogate mothers, the terms and conditions to be followed during the surrogate phase, the amount to be paid and violation of surrogate mothers' rights, if any, the nature of support system required, etc, were gathered through personal interviews. Focus group discussion was also organized, in order to get the collective opinions of surrogate mothers, but it was not possible due to some disturbance faced by the Members. With the limited number of responses, the study had to confine itself to present the case studies and some basic analysis of their socio economic profiles.

Based on the thirty three interviews with surrogate mothers, characteristics, issues, challenges, prospects of the participants were analyzed. The data collected were interpreted and on the ethical considerations involved were discussed with the participants and they are presented from a gender perspective.

Research Methods and Techniques Used

The study was conducted in Tamil Nadu, taking Chennai, which is the Capital City of Tamil Nadu, where the surrogacy was concentrated. The Researcher selected Chennai Metropolitian City as the study area because Chennai has more number of fertility centers and surrogate mothers, in come to Chennai, large numbers to be away from their relatives for 10 months. Besides, each and every day, fertility center is proliferating. Researcher tried to get details from all fertility centers, practicing surrogacy and some of the fertility centers were reluctant to reveal the data. Many practicing clinics informed that they did not practice surrogacy. But the fact is that Chennai City had turned into a heaven, for foreign commissioning parents, due to the low cost of reproductive technologies and economical surrogate mothers. The Globalization Era has changed Chennai City into a center of Reproductive Tourism, for the overseas citizens. Having chosen Chennai, the data were collected, using qualitative methods. Applying narrative approach were gathered data with unstructured interview questions. Personal interviews were used to get the data from the participants. In addition, various stakeholders, connected with Surrogacy, were also interviewed, including gynecologists, agents and researchers, experts in the field etc. in order to get a holistic idea and perception on surrogacy. The present study analyzes the experience, issues and challenges of participants, who were often the victims of social oppression as evidenced in several studies. Experiences of different surrogates were analyzed with feminist lens. The Researcher relied on collaborative dialogue with the participants, for further understanding. Collaborative dialogue emerged during interviews between the Researcher and participants, which added to the narrative interpretation, focusing on relationships. In other cases, it focused on social and cultural environment, that shaped the life events of surrogates and the meaning attached to them. Interpreting the participants' narratives, key themes, plots, and discourses were the focal points for the analysis. Hence a mixed method was adopted, to collect data for analysis.

Data Collection and Challenges

Both primary and secondary data were collected for the study. Necessary secondary Data were collected from various research reports, relevant books, journals, internet sources, published and unpublished articles, newspapers, dissertations, web sources etc. Necessary secondary data, from various fertility clinics and the surrogate mother associations, were collected, from the Website and Print Media sources.

As already mentioned the Researcher tried to meet Doctors, Agents or Intermediates, Associations and the Surrogate Mothers, Mediators, who were involved in the surrogacy process. But they hesitated to meet the Researcher. After prolonged struggle and rapport building, the Researcher got the consent to meet the Associations and convinced them by explaining the need. Later the Researcher got the opportunity to interview the surrogates. Six months of continuous request from the Researcher, with some recommendations from various sources, bona fide certificates and the separate official letter to the Association, by the Research Supervisor, the Association arranged for a meeting with three surrogate mothers for the pilot study. It was very tough to identify the surrogate mothers, without the Associations' help because the surrogate mothers were not willing to reveal their identity to anyone. With the help of them, pilot study was carried out. In the pilot study, the surrogate mothers were asked about their socio-economic details, current living status, details about marriage, type of marriage, dowry, alcoholic behavior of husband, socio- economic and health issues due to surrogacy and their complete delivery experience in terms of physical, mental health, nutritional, sexual, and economic, its challenges and impact on self and children, reactions, legal action etc.

The Researcher had to travel all over Tamil Nadu, to contact various fertility centers in person. But the Researcher did not get any positive response from the IVF centers and a few of them had replied that they were not interested. The Researcher could not meet the family members of the surrogates, commissioning couples, doctors and the brokers or intermediates. There was difficulty in identifying the agents or intermediates Associations, who were working with surrogates and to meet them for getting details of the surrogate mothers, in Chennai. Even after getting the consent, from them the surrogates hesitated to disclose their identity to the Researcher. Basically, the surrogate mothers were economically poor and they had to work for their survival. The

Researcher struggled to organize the surrogate mothers for a Focus Group Discussion. No one was willing to participate and share their experience in the group. Also, they did not like to show their identity to other surrogate mothers. It was difficult to find separate space for interviewing and share their experiences and there was no privacy to express their views. Majority interviews were done on the roadside and other public places. The Researcher found it difficult, to convey and receive the message in colloquial language, which could help them to understand the concept, related to surrogacy.

For the primary data, a pre tested unstructured interview schedule was prepared to collect the information, from the surrogate mothers. The schedule was prepared in English. In addition to the qualitative means of data collection, voice recording of the interview was also collected to substantiate the findings. In this research, audiotape was used for data collection, with informed consent. The advantage of this data collection was that the entire data collection could be made by the Researcher, by concentrating on interviewing the participants. Data on demographic and economic profile, factors associated with surrogacy, mental health problems, surrogate mother perceptions on surrogacy in the perspectives of gender etc were gathered. The participants, who gave their permission for their participation in this study, were promised confidentiality. This was necessary because many participants agreed to be surrogates, without their family's knowledge and consent. As a result of assurance by the Researcher, the participants were willing to share their experience. The data, collected through voice recording, field notes and case studies, were thoroughly documented, categorized and presented under various conceptual frameworks.

Sampling Method

The study had used non-probability sampling as it was not possible to get the universe and use random sampling, representative of the universe. One of the non-probability sampling methods, namely, snow ball sampling (or chain sampling, chain-referral sampling, referral sampling) was used for collection of data.

While the Researcher approached the fertility clinics and associations, through e-mail and in person, to get the list of surrogate mothers, they were reluctant to reveal the identity of surrogate mothers. The Researcher was unable to identify the universe of surrogate mothers from the fertility clinic, in spite of repeated visits, requests and persuasion. Hence the Researcher selected a sample of 33 participants of married women who were willing to participate and give information. The 33 women sample was arrived at by applying the snow ball method, to study the very sensitive issues and challenges involved in commercial surrogacy. Of the 33 participant's interviews, 30 participants successfully delivered the surrogate babies, as informed by the participants. Two surrogate mothers were pregnant during the process of data collection and one woman reported failure.

Data Collection and the Interview Process

Researcher interviewed thirty three participants for this in depth narrative study. Interviews took place from January to August 2015. Interviews were about one and a half hours to two hours in duration. The interviews were non structured, emergent conversations, about the experience of commercial surrogacy. Researcher conducted open-ended interviews with participants, using the questions outlined in appendix, as a starting point. According to Weiss (1994), majority of significant events in people's lives, can become known to others only through interview as interviewing opens a window to the past and rescues events that would otherwise be lost. During the interviews, participants asked about the Researcher's experience, which facilitated in depth conversation. From the interview session, the Researcher arranged the answers, under themes, for the analysis. The data collection process served as a dialogue process not only to collect data but also for sharing the experiences of the surrogate mothers and creating awareness on the rights of the surrogate mothers. Hence the Researcher performed the twin action of collecting data and imparting knowledge on surrogacy rights. It is to be noted that with all care, the data collection was over and the analysis was completed in 2017, as the Researcher felt that sufficient data could be supplemented by including some more samples and also conducting interviews with other associated stakeholders. By then Government of India came out with the Draft New Surrogacy Bill 2016 which could cover many of the recommendations that the present research has polished. Hence an attempt has further been made to collect data from doctors, fertility clinics etc, to have their views and opinions on the Draft New ART bill 2016 and they were analyzed and presented.

Tools of Analysis

The Researcher had employed simple statistical tools, such as percentage analysis to analyze socio economic profile of the participants. Percentage Analysis involves computing measures of variables, selected for the study and its findings will give easy interpretation to the readers. During the analysis of the narratives on commercial surrogacy, the Researcher paid particular attention to how the participants described their familial relationships, the meaning of their experiences, their choice of language, and how these accounts fit into various discourses. The Researcher's approach to analysis, involved in depth understanding of the themes, the narratives, to assist with meaning making and understanding the experiences of participants, their family relationships, and the social context. The Researcher chose issues and topics, that had relevance for the research literature and the broader community. Themes related to the overarching research question, were highlighted. Themes, connected to how surrogate mothers make meaning out of the experience of commercial surrogacy, were also included. In addition, themes were interpreted by focusing closely on language, in each of the stories while using a temporal lens. This means following and exploring themes, in a chronological order, from the start to the end of a story. Winter and Daniluk (2004) explain that, "we identified themes for each personal narrative and followed these themes throughout the story as they evolved from beginning to end." Prior to writing the narrative, the Researcher read all the transcripts, for themes with similarities, points of difference, insights, and how the women made sense of these various topics, within the commercial surrogacy. Further, the Researcher paid particular attention to how the story was being told, by each commercial surrogate, during the interview. In essence, the Researcher immersed herself in the transcripts, through multiple readings and sat with the interview dialogue, cognitively and emotionally, before starting to write.

The Researcher approached the task of writing the narratives, by organizing transcripts in the chronological, order in which the interviews occurred. In addition, the Researcher wrote the stories of surrogates, by using the temporal ordering of their experiences. However, the development of the narrative, through the emergent non structured interview process, often took an unexpected path. For example, a discussion about the challenges of surrogacy versus commercial surrogacy, unfolded in the emergent interview process, even though this was not a question in the schedule. Also, the Researcher wrote the narratives, by identifying similar themes that emerged most often and evolved within and across the thirty three stories. Therefore, the Researcher included these excerpts, in the surrogate's words, from the personal narratives. According to Reissman (1993), the challenge is to identify similarities, across the moments, into an aggregate, a summation.

In bringing out the implicit and explicit themes, within and across stories, words from the surrogates were directly used. Fear of keeping the surrogate baby is an example of an implicit theme and medical practice and procedures is an illustration of an explicit theme that was identified within and across the stories. Headings, for the thematic segments, in the personal narratives were arranged from the surrogate's own words. During the conversation, different points of view, about their experiences, were revealed, which became a site for analysis. For example, one surrogate decided to feed her breast milk to her surrogate baby, which was a point of difference amongst the thirty three stories. In the individual narratives, an analysis of a main emergent theme from the personal story was undertaken and an overarching heading was created. This piece of analysis was included within the personal narratives and each individual story provided a breakdown of some main themes, in commercial surrogacy, as they emerged.

The Researcher identified significant themes, under the categories of emotions or feelings, points of resonance, distinctions, relationships and politics. The interview transcripts were re-constructed into individual narratives. In analyzing the themes, the Researcher drew on social constructionist references (Gergen & Gergen, 2004; Thomson, 2005). Subsequently, the Researcher looked across the transcripts and focused on the influence of various discourses, for the analysis, from a meta-perspective. An in depth inquiry of the unspoken rules, within and across the narratives, was undertaken. This analysis, with narrative construction of the stories and how these stories were influenced by dominant societal discourses was also brought out.

Scope of the Study

Given the challenges in conducting the study, the contribution would be immensely helpful to understand the socio, economic, cultural and political questions behind surrogacy. It is an exploratory research, where the issues of surrogate mothers on the 'Rights' perspective, was also analyzed. Since the willingness of the surrogate mothers is quite restrictive and limited in sharing the information, the study would focus on narratives and employed the feminist research methodology, applying largely the qualitative techniques. An intensive analysis with narratives, anecdotes, sharing of experiences, voices, etc, covered the research problem, thereby offered a universal applicability and generalization. The study heavily relied upon the Association working for the welfare the surrogates and hence it was not able to cover the non-members of the Association, which specifies the need for further research by encompassing such groups. It is a policy research where the conclusions and recommendations, would directly help the policy makers. However, there are overlapping findings and recommendations, after the recent surrogacy draft bill 2016.

Limitations of the Study

The study was conducted in Chennai, the Metro City of Tamil Nadu, as Chennai had the largest number of Fertility Centres in Tamil Nadu, at the time of conducting the study. The study was conducted only among the women, who acted as surrogate mothers, with inputs from the Associations. Also in this study, the commercial surrogacy alone was examined by the Researcher because commercial surrogacy was an emerging issue in the global scenario. The Researcher could identify only limited participants. During data collection, there was conflict between the clinic and the surrogate, regarding remuneration. That is, fertility clinic refused to pay the amount promised after the surrogate delivered twins. The sample was provided by the Association, working for safeguarding the rights of surrogate mothers. There may be other surrogate mothers outside the Associations, who might have to narrative different experiences and violation of rights. These things could not be brought out in this study. Moreover, it was assumed that the issues faced by the surrogates of the sample, given by Association, would reflect the universal issues faced by surrogate mothers and as such, the findings of the study may be relied upon.

Chapter VI

Analysis of Data and Interpretation

The analysis of data has been presented in three sections. Socio economic profile and analysis of the motivating and negotiating factors, experience and issues and consequences faced by the participants during and after surrogacy and participants' perception about commercial surrogacy and gender. Qualitative and Oral Testimony Interpretation was widely used, to analyze the data, with the help of various conceptual and theoretical frameworks. To some extent, the study qualified the qualitative data, by presenting the findings with accuracy (percentage).

Surrogate motherhood is a complex technical process, defined in different ways by different people. Before performing the analysis of the collected data in Chennai, Tamil Nadu, it is important to decide the exact definition of surrogacy, in different contexts. Online Cambridge Dictionary defines that "*a woman who has a baby for another person, who is unable to become pregnant or have a baby*." The ART Regulation Bill, 2010 defines the surrogate mother as "*a woman who is a citizen of India and is resident in India, who agrees to have an embryo generated from the sperm of a man who is not her husband and the oocyte of another woman, implanted in her to carry the pregnancy to viability and deliver the child to the couple or the individual who had asked for surrogacy*."

Profile of Surrogate

Women, who undergo surrogacy, are usually very poor in economic condition. To earn money, they accept to relinquish the baby for another couple because for surrogates, it will take several years to earn that kind of money. Also, these women were not able to get regular earnings, due to lack of job opportunities and education. While these women were engaged in household and casual works, most of the time these women migrated to some other areas, to get a better job and income. Hence these women were living in the slum areas, due to lack of housing facilities and the high cost of living. The lower middle class and married women agreed to surrogacy, to earn huge money at quick time, in order to improve their status in the society, to fulfill the basic needs of the family, for children's education, medical expenditure, build or buy house, start business for their illiterate drunken life partner. Due to poverty, lots of women wished to serve as a surrogate. The Indian commissioning parents realized their economic condition and agreed to pay a huge amount to the surrogate. Compared to Indians, the foreign couples paid more. For procreation, fixing of amount was based on the skin tone, education and caste. These factors did play a major role in selecting a surrogate.

Having discussed the theoretical framework, the concepts, the review of related studies in the earlier Chapters, the data analysis and discussion would be presented in this Chapter. The socio economic background of participants is presented in Section I.

Socio economic characteristics are an important component of social science research. It is the socio economic background, which influences the nature of participation of women outside the household and determines the status of women. Hence the socio economic background of the participants was analyzed, to find out status of the women, who served as surrogate, in terms of education, age, standard of living, nature of residence, the community, the religion, nature of work, etc. The economic background was analyzed, to find out the total income earned, the number of dependents, the poverty level, the challenges for livelihood needs, number of members earning in a household etc. The economic background was also used to find out the monthly expenditure on alcohol by the partner of the surrogate. The socio demographic and economic background variables certainly exercised high degree of influence on the living standard, skills obtained and the income they commanded.

Section I: - Socio Economic Profile

Socio economic profile of the participants was analyzed in order to examine whether there was association between the socio economic background of the surrogate mothers and their decision in favour of surrogacy. It is often claimed and evidenced in several research studies that the surrogate mothers do not come forward to serve as surrogate, once the opportunity is open to them. They are persuaded by various agents including the brokers, friends and peer, the existing surrogates in the field, the doctors etc. which ultimately either motivate or push the women to offer their consent to serve as surrogate. There are myths that surrogate come to such a profession to earn more. Having given their consent surrogate mothers take lots of effects, to convince the family members who initially resist such offers. Hence the socio economic background was analyzed.

Socio economic profile included the age, education, size of the family, income and other socio economic aspects of participants.

Age

Age is an important factor, to understand which age groups of people are mostly selected by the commissioning parents and the physician. Age is also an essential factor in identifying and documenting the risks, issues and challenges, faced by surrogates. Age plays a key factor in determining women's participation in family, society and market and hence it was analyzed.

Table 1: Distribution of the Participants by Age

S. No	Age	Frequency	Percentage
1.	21 to 25	04	12.12
2.	26 to 30	13	39.4
3.	31 to 35	06	18.18
4.	36 to 40	08	24.24
5.	Above 40	02	6.06
	Total	**33**	**100**

(Source: Primary data compiled from data collection), Average 31years

The age wise distribution of the participants indicated that majority of surrogate mothers, were aged between 26 to 40 years. This might be the ideal age group to serve as surrogates. There are several terms and conditions, that surrogate mothers are compelled to follow. Majority of surrogates, in this study, as well as in other studies, reported that they had two children.

In fact, women were brought to doctors, as egg donors, which was accepted by women after a lot of persuasion. Later when the institution restricts the number of times a woman is allowed to donate egg, they are slowly persuaded by their own friends and peers, to serve as surrogates. If the conditions were fine, like they did not undergo cesarean more than once, etc, they were taken to doctor for fitness and other tests. Finally, the market succeeds in making the women, to accept to serve as surrogates. Age and physical condition determine whether a woman can serve as surrogate. In other words, women with suitable age, who are healthy enough to bear a child, alone are preferred and other women, through interested in serving as surrogates, may not be considered

Table 1 reveals the Distribution of the Participants by their Age. More than one third of the participants were between the age of 26 to 30, followed by less than one fourth in the age group of 36 to 40. It shows that majority of participant surrogates were in the group of age 26 to 30, after delivering at least their own child. It was inferred that at this age, surrogates were healthy and suitable for being surrogates. Some studies have mentioned that women, at this appropriate age have been mostly targeted by the agents, to undergo surrogacy, for commercial purposes.

Marital Status

According to the Indian Draft ART Bill, married women alone can serve as a surrogates and it is an offence, if unmarried women were to offer to be surrogate mothers. Now the Draft new Bill has further restricted the eligibility to serve as surrogate mothers.

Table 2: Distribution of the Participants by Marital Status

S. No	Marital Status	Frequency	Percentage
1.	Married -Living with Husbands	28	84.8
2.	Divorced - Living with Parents	01	3.0
3.	Widow - Living with Parents	04	12.1
	Total	**33**	**100.0**

(Source: Primary data compiled from data collection)

It is evidenced in the study that maximum number of married women living with husbands, had opted for surrogacy, though there were a few widows and divorced also in the sample. In the given cultural context, it is quite surprising to see that widows had opted for surrogacy, breaking the taboos and stigma. On the other hand, it also reveals how vulnerable the sample widows were because they were willing to hire out their uterus for want of money. In such cases, the chances of being exploited, will be relatively more, by various agencies involved in the process of surrogacy. There is perception that single women, widow, separated or divorced, do not have families and since there is no one to support (relatives, wards, kith and kins) they are easily exploited by the market. In this study, it is observed that more than four-fifths of the participants, were able to convince their husbands or parents, to give consent to serve as surrogates. It is good that women were able to convince which may be attributed to their increasing bargaining power at home, due to the huge money they were likely to collect at the end of surrogacy. The unmarried women, if denied, cannot serve as surrogates. On the one hand, it may be perceived as denial of women's rights as the women who wish to serve as surrogates are denied the right to perform. The cultural norms prevent unmarried women to opt for surrogacy as a profession as it would give rise to commercial market, which would exploit women further. This is quite inapplicable to Indian social norms where the married women taking to surrogacy, may not carry a stigma while unmarried, in a patriarchal perspective is to be chaste before marriage, where the society may see unmarried giving birth to a child, is deemed to have lost her chastity.

It is inferred that 84.8 percent of the participants of the study were engaged in surrogacy, with the consent of husbands or parents. It is good that women did participate in decision making and they were capable of convincing their husbands. It was accepted by the husband because it was the process which provided economic support to the family. Some unmarried women wish to take it to surrogacy as an employment opportunity. According to the ART Bill, married women alone can practice surrogacy. This is a kind of denial of women rights. Women rights are human rights and the ART bill indirectly and strongly shows patriarchal bias. From the gender perspective itself, dignity and individuality of women is denied. From the feminist perspective, all women (single, married, widow) should be given equal employment opportunity, to get some employment. But ART Bill denies employment opportunity and creates inequality between women. On the other hand, if widows and separated women are allowed to practice surrogacy why not single women. Since unmarried women are not allowed to practice surrogacy, patriarchy strongly denies women their reproductive rights and thus it also interferes with the right of women to decide on surrogacy.

Israel permits commercial surrogacy but the *Embryo Carrying Agreements Law* (1996) stipulates that a "severe effort" be made, to permit only unmarried women to serve as surrogates, because it is reasoned that allowing married women to serve as surrogates would violate culturally prescribed definitions and norms about kinship, the status of the child that is born, and family (Kahn 2000). The living status of surrogate indirectly portrays that without any support, women cannot be surrogates. Also, it is assumed that women should be "dependent." The woman, who is living with husband, could be forced to become a surrogate. There are different kinds of employment opportunity but due to lack of financial support, women are forced to become surrogates, to meet their financial needs. Also, divorced, widowed and single women became surrogates due to lack of employment opportunity. In the end, it is explicit that double the time women are being exploited.

Type of Marriage

Marital status, in this study, was not limited to only the marital status but also the type of marriage, marital duration, the choice of marriage etc. In the Indian culture, marriage is the major reason for procreation and it is not accepted before marriage. In this study, all the participants were married. Nearly 40 percent of participants married in their childhood i.e. marriage happened before 18 years of age.

Table 3: Distribution of the Participants by Type of Marriage

S. No	Type of Marriage	Frequency	Percentage
1.	Choice	19	57.6
2.	Arranged	14	42.4
	Total	**33**	**100.0**

(Source: Primary data compiled from data collection)

It is quite surprising to observe that nearly 50% to 60% of the participants, in this study, had their marriage by choice. It needs to be analyzed whether such choice had given them the freedom and liberty, to convince their partners to permit surrogacy. In the field it was observed from the conversation with the participants that majority of couples, who had opted for marriage by choice, were not accepted by the parents for reasons well known. Non- acceptance of such marriage by parents, inevitability push the couple to manage the life on their own. Majority of the couples were forced to prove to their parents that they had selected right partners and also they wanted to prove to their parents that they were leading a peaceful life. Issues to be discussed in the context of nature of marriage, in the Indian context, are the caste based challenges. The couple, who wish to choose their partners, are often prohibited to do so. In certain cases, if the partner is from upper caste for the girl it is accepted but the reverse is not permitted. A scheduled caste (women in the lowest level in caste hierarchy) man cannot opt to marry an upper caste woman. If it happens, it could lead to honour killings, which is the new and emerging form of crime. The man, from scheduled caste, will be murdered by the girl's family. In this study, the surrogate women were quite progressive not only to choose their partners by their choice but they also left their parents, in the hope that they could lead their lives on their own. But the challenge before the sample surrogate participants, was that they had to prove that by going with the partner of their, choice they were leading a peaceful life. Hence the women had to opt for surrogacy, to earn their livelihood. The partner also extends has support.

There were two kinds of pressure, for the surrogate mothers. If the partner chosen was able to bring adequate income to meet the household expenses, it will be managed. But if they failed to earn and at the same time, if the family were to extend with children, the couple might resort to indiscriminate borrowing. When the debt gets accumulated, they are forced to earn to pay back the same. In such a context, surrogacy serves as an attractive option. The agents or the brokers rightly identify such vulnerable couples, to serve as surrogate and the economically challenged couples, easily yield to such offer, without any dispute. But it does not mean that couples, who made their marriage by choice, are all pushed to surrogacy. Another major factor is that in rural areas, early marriage is common for fear of 'eloping' of adolescent girls and that too, with lower caste guys. Such early marriage further gives scope for serving as surrogate mother. Once again it does not mean that all inter-caste marriages are being questioned by parents or honor killing takes place everywhere. However, even if they are only a few cases, it should not be encouraged to continue. One should move towards casteless society. In fact, surrogacy does not discriminate, on the basis of caste. Caste has never been an issue in the market. Most of the couples, who opted for own choice marriages, were not accepted in the family, due to cultural conflicts. Since they were excluded from their family, they could not get sufficient income for their livelihood. Hence they opted for surrogacy and they took their decision on their own and there was no need to consult anyone. Maximum number of couples had to prove to their parents that they had selected the right partners and also they wanted to prove to their parents that they were well off. In other words, participants wanted to prove that by making a choice of their marriage on their own, did not suffer in any way and their decision was right though the participants were struggling for their economic needs. Under such circumstances, the participants had to convince their husbands to become surrogates. Child marriage was also a major kind of forced abuse on women and the child marriage indirectly led to surrogacy. Due to social stigma, majority of women could not choose their partners. After marriage, if partners had insufficient income, they cannot blame their parents. Hence these women were forced to do surrogacy, due to financial constraints. Traditional reproductive process forced women, to do stereotyped gender role, as an economic activity. In other words, economic factor reinforced women to go for traditional reproductive work.

Age at Marriage

"Age at marriage" for women is important for a happy married life. Age at marriage can differ from one religion to another religion. According to the Indian law, minimum age, at marriage, should be 18. It is scientifically proved that girls at this age, are ready for the reproductive process. Theoretically, it may be below 18, due to social vulnerability and insecurity of girls.

Table 4: Distribution of the Participants by Age at Marriage

S. No	Age at Marriage	Frequency	Percentage
1.	Below 18	14	42.42
2.	Exact 18	07	21.21
3.	Above 18	12	36.4
	Total	**33**	**100**

(Source: Primary data compiled from data collection), Average: 17.8years

Table 4 depicts the distribution of the participants by Age at Marriage. Little bit lesser than an average of participants (42%), married before they reached the minimum age for marriage. This shows that in the Indian context, men prefer a girl, who is at least 10 years younger to them or a girl who had just attained puberty. There is a belief that if men and women belong to the same age group, they cannot lead happy married life in their sexual and social spheres. In the patriarchal society, parents consider girls as burden. There is a pressure to maintain virginity and safeguard their daughters from various abuses and to maintain their morality which forced them to get into early marriage.

It is inferred that nearly 50% of participants had married before the legal age. It is one of the major reasons, for women to agree to undergo surrogacy. Infatuation, love affairs, social security, ill health of parent, which prompted girls to get into early marriage.

There is an association between early marriage and susceptible to ill health, poverty, children's welfare and ultimately, to economic vulnerability. Early marriage has been forced due to two reasons: to protect the girls from physical and sexual violence due to the social expectation that women must be virgin. Second the lack of maturity, both in physical and mental health spheres, assuming that young girls will become mothers and assume the responsibility of bringing up their children with a better status than theirs. All these forced women to get some employment and these women go out for work. In their work contact, women are motivated by their peers, to donate egg in the first instance, which brings a lump sum amount and later, persuaded by the agents and peers to get into surrogacy, which assures a considerable income in lakhs, which is often a dream for poor income households. As it has emerged that majority of the surrogate mothers, who opted for marriage by choice, were not able to get their support from parents as these couples might have eloped and failed to get their family support. In such a situation, if they do not get a decent job, it is highly challenging to meet their livelihood needs and hence resort to jobs, which will assure them adequate wage, to pay off their debts. Under such circumstances, surrogacy arrangements come as a blessing and they easily accept the agreement, without any inhibition or convinced to accept, to serve as surrogate mothers.

Age at Surrogacy

According to the Indian draft ART Bill, no woman, less than 21 years and over 35 years of age, shall be eligible to act as a surrogate mother. Also, no woman shall act as a surrogate for more than five successful live births in her life, including her own children. Abortion, still born, etc must also be included as it does count as pregnancy. But in reality, due to the livelihood needs to be met, and they are forced to hire out their uterus, by violating those age norms or the number of children born already.

Table 5: Distribution of Participants by Age at Surrogacy

S. No	Age at Surrogacy	Frequency	Percentage
1.	20-25	07	21.21
2.	26-30	12	36.36
3.	31-35	13	39.4
4.	Above 35	01	03.03
	Total	**33**	**100**

(Source: Primary data compiled from data collection), Average: 28.8years

Table 5 presents the distribution of participants, by Surrogacy Age. In order to understand, which age group of women were targeted for surrogacy, the age of participants was analyzed. The age, at surrogacy data, clearly informs that it very much happened in the reproductive age and in this study, the maximum age was found in the range of 26 to 35. It further informs that these women, by this age, would have completed their role as mothers and they were available to donate their uterus for surrogacy. However, it cannot be denied that 26 is a very young age where these women were pushed to serve as surrogate mothers.

The age between 21 and 3, is considered the right age for surrogacy because at this age, women's reproductive health organs are good and they are in a state of active procreation. It is further understood that surrogate mothers had both positive and negative experiences to report in their personal life. Due to economic crisis, the patriarchal society pushed them to take a decision on their own and avail the support from their partners, to undergo surrogacy. The right age for fertility process was also the right age to take decision, for their survival. The partners were reluctant in the beginning due to the taboos associated with the surrogacy as it is sometimes deemed to be equivalent to prostitution. But they consented on the grounds that it brought a lump sum amount. This is mutually beneficial. The money helped the donor economically and on the other, it is sort of helping the infertile couples to have their children, which is a service. In a way, it is a social enterprise where the service component appears to be inbuilt. Age at surrogacy is an important variable in the surrogate research, where women just have to donate the uterus and hence even if the woman is above 60, can offer surrogacy and it had happened also. But the feminist critique would like to know whether the body of women be put into experiment, even at the age of 60, for the sake of getting a child? These criticisms have never been answered in the society, which is highly patriarchal.

Educational Status

Education is a tool of social change. There is a direct relationship between education and surrogacy. Education plays an important role, especially in the development of women. It helps to increase their self-esteem and dignity and helps them to be aware of their rights and to fight against injustice, abuse and inequality, exhibited towards them. Education also helps women to land honest jobs and lead to financial liberty, decision-making capacity, self-confidence and independence and thus improves their socio economic and political status. Illiteracy prevents women from getting a decent career.

Table 6: Distribution of the Participants by Education

S. No	Education	Frequency	Percentage
1.	Illiterate	01	3.0
2.	Literate to just sign	09	27.3

S. No	Education	Frequency	Percentage
3.	Primary	08	24.2
4.	Middle	05	15.2
5.	High School	08	24.2
6.	Diploma	01	3.0
7.	Degree	01	3.0
	Total	**33**	**100.0**

(Source: Primary data compiled from data collection), Average: 2years

Table 6 exhibits the distribution of participants by Education. It is observed that vast majority of participants (94%) did not complete their schooling. To earn money, women opted for surrogacy, without knowing the future impact of surrogacy on their health particularly the post natal issues for which there are no arrangements. It is against nature because the child is deprived of breast feeding and the mother must also has to go through the post natal process, without any issues. But the surrogate mothers are forced to sever relationship with the child, where both the child and the mother of the the baby are forced to go against law of nature and suffer. After surrogacy, the child will be denied breast feeding and equally worst is for the surrogate mother too, where she suffers which post natal heaviness in breast, due to breast milk clotting, which would sometimes cause dreadful pain. In other words, women's health and the child's health are least cared for during surrogacy. That is why in the Sangam Literature era, in epics like Ramanayana and Mahabharatha, there were arrangements that the housemaids will also be pregnant and in the case of the Queen failing to breast feed, the servant community must make themselves available to offer themselves to feed the child. Today with medical advancement, medicines are given to overcome post natal issues but still it may not address fully the nature of pain. Feminists point to such pains and health issues, where the market does not have any arrangements as the market always functions without any ethics and human considerations.

Education, as a variable, played a major role in understanding the technological procedure of ARTs, though majority of them had not crossed school education. It is true that education must create awareness, but due to lack of education and awareness woman underwent surrogacy. Meanwhile, lack of education prevented women from getting more employment opportunity.

Religion

Religion also plays a major role in the surrogacy process. Many cases of the discrimination happened on the basis of religion. Almost all the participants belonged to Hindu religion. While Hindus and Christians accepted the surrogacy, Islam prohibited surrogacy. Religion is so effective, but unfortunately, it is in the hands of fundamentalists, who concentrate more on maintaining regressive norms than on progressive ones. The future generations must privilege alternative perspectives to rewrite and deconstruct the social norms which are unfavorable to one gender, which constitutes social injustice.

Table 7: Distribution of the Participants by Religion

S. No	Religion	Frequency	Percentage
1.	Hindu	26	78.8
2.	Muslim	03	9.1
3.	Christian	04	12.1
	Total	**33**	**100.0**

(Source: Primary data compiled from data collection)

Though Islam prohibits surrogacy, the economic needs and livelihood challenges might have forced the Islamic women to accept surrogacy as an employment opportunity and served as surrogate. It is quite unacceptable to Islam but still around one tenth of the participants did hire out their uterus, which only indicated that when hunger strikes, the norms imposed by culture, religion etc are ignored. This is the base for the commercial sex workers also. No woman is born to accept sex work as the profession, but when it comes to meeting the livelihood demands women are forced to take a decision to go as sex workers. Same logic applies that though Islam does prohibit, the households who really struggle to survive, do not mind the religious or socio cultural taboos and resort to some kind of work, to feed the members of the family and help them to survive.

According to Hindu mythology, surrogacy is accepted and more women undergo surrogacy but in Islam (sarath), practicing surrogacy is strictly prohibited. Even though some muslim women practised surrogacy, it was only due to abject poverty and economic needs. Thus muslim women were pushed into surrogacy. In such a scenario, if the Islamic society came to know about surrogate muslim women, their family was socially ostracised from the Islamic society.

Region

Data on the region, from which the surrogate mothers hailed were taken to find out the region of origin and operational areas from which surrogates were drawn from. Region here refers to rural or urban. Such information, about the location of residence, was collected primarily to explore access to surrogacy services and variations in the cost of living and levels of affordability. The data, about the residence or location of participants, enabled the mapping of the journey undertaken by them, to access the surrogacy process, the struggles to leave home to stay for a year away, the nature of convincing to be done to members of the family, children etc. This was useful to analyze the nature and extent of movement by users, who may have left their places of residence, to act as a surrogate. It would also reveal the penetration of medical tourism market, which, had to hunt for surrogate mothers because it is not visible and mainstreamed.

Region also plays a key factor in the surrogacy process. The majority of participants were from urban area, which was around the metropolitan city. There is a belief that compared to rural area, career opportunities were more in the city. But if they have lots of job opportunities in the city, then why do women decide to act as surrogates?

Table 8: Distribution of the Participants by Region

S. No	Region	Frequency	Percentage
1	Urban	31	93.9
2	Rural	01	3.0
3	Semi urban	01	3.0
	Total	**33**	**100.0**

(Source: Primary data compiled from data collection)

Table 8 demonstrates the distribution of participants by Region. The data also indicated that compared to rural, the urban women knew about the surrogacy procedure. It is inferred that vast majority of participants (94%) were from urban areas, which shows that urbanized people did have knowledge of surrogacy and its process. In our society, pregnancy, child bearing, rearing and caring are respectable work for women. Though surrogacy is stigmatized, poverty compelled more urban women to break the social stigma in order to overcome their poverty. Compared to rural women, feminization of poverty prompts more urban women, to break the stereotyped role and use the opportunity to earn instant money, through surrogacy, for their survival. It is not surprising that urban women were more in surrogate markets as they get more information than the rural areas. But the fact remains that these urban women were from the lower rung of the society.

Caste

Social institutions play and determine the activities and shape the social norms. Among the social institutions, caste has become an inevitable factor in the socio-economic analysis. In this study, caste played a major role in the selection of surrogates. There was preference for a certain caste, due to reasons of reproductive health and issues of payment.

Table 9: Distribution of the Participants by Caste

S. No	Caste	Frequency	Percentage
1	SC	16	48.5
2	BC	08	24.2
3	MBC	08	24.2
4	FC	01	3.0
	Total	**33**	**100.0**

(Source: Primary data compiled from data collection)

The economically and socially vulnerable communities, like Dalits and Other Backward castes, have been represented by more than three fourths of the surrogate mothers in this sample, as evidenced in the Table. The high percentage of SC participants indicated that surrogates were mostly from Scheduled Caste because they were recognized as socially and economically vulnerable, less educated, not much informed about the political economy of surrogacy. Of these factors, their economic vulnerability was the key factors, which forced them to accept low fixation of surrogate charges. It is inferred that though women from all castes were involved in surrogacy, preference was given to forward castes the highest in the hierarchy of the castes in Indian caste system, as the choice of intended parents. Insufficiency of surrogates in the specific caste, compelled them to make a choice, with the low caste in the hierarchy of Castes, namely the Scheduled Caste (Dalit), who constitute the maximum in this sample, who were not only socially backward but also economically and they were willing to be surrogate mothers, with relatively low cost. Caste preferences were not observed in this study but religion based participants were selected by the commissioning parents fair beautiful women were preferred by the intended parents (The Hindu, 2012).

Generally, upper caste surrogates were preferred. But in this study, in the process of agreement to go for surrogacy, intended parents did not know about the caste background of surrogates. Due to the economical deprivation, more women underwent surrogacy. Meanwhile, scheduled caste women were considered to be rendering the cheapest labour in the surrogate market. Thanks to the confidentiality being maintained in the surrogate market, the caste background is not revealed to the commissioning parents. Caste and Gender intersect and here the women are needed to make the infertility issue addressed but at the same time, designer babies like higher caste, fair are babies privileged. Feminist criticize that women are already put to vulnerability, as surrogates and it is all the more challenging for the scheduled caste women. Women are doubly disadvantaged when they are from lowest caste hierarchy. It is to be argued that the infertile couples are suffering from childlessness and it is enough if they get a baby, irrespective of which background the surrogate mothers belong to, but unfortunately the caste system remains an inevitable variable to be reckoned.

Type of Family

The joint family system is almost dwindling and nuclear families are on the increase. It is a key factor for understanding the level of decision making of women in surrogacy. Due to globalization and generational gap, joint family is on its way out and the emergence of nuclear families system is widespread, in both rural and urban settings. Due to changed circumstances, women are forced to earn their livelihood and fulfil their basic needs. Participant families were categorized into joint and nuclear family. The nuclear family included duos and their children only while joint family included couple, children, grandparents and all other relatives living with them, under one roof.

Table 10: Distribution of the Participants by Family Type

S. No	Family Type	Frequency	Percentage
1.	Nuclear	29	87.9
2.	Joint	04	12.1
	Total	**33**	**100.0**

(Source: Primary data compiled from data collection)

The distribution of the participants, by nature of family, revealed the macro trend of in India nuclear family is on the increase. The family structure is the main reason for the participants to take the decision regarding surrogacy. In a nuclear family, the couple can take their own decision. All the participants belonged to male-headed households. It clearly reveals that participants became surrogates, with their partner's permission, to support their family economically. But in the joint family, social stigma and the exclusion criteria compelled the couple, to conceal the entire process of surrogacy. In addition, it was also observed that couples, in order to tackle immediate financial crisis in the family, took this decision, without consulting or seeking the approval of their elders in the family. At the same time, participants, living in a joint family, enjoyed the family's financial support. But the participants, from nuclear family, did not enjoy such support and hence they were compelled to undergo surrogacy. It is also to be mentioned that the joint families help in taking care of children when the women are away for surrogacy and such support systems are required for peaceful delivery of the babies.

Number of Children

According to the proposed ART Bill, women opting for surrogacy, should not have children exceeding five, including the surrogacy delivery. Total number of children of participants has been taken, to identify whether the participants had children of their own and it would be helpful to identify whether the physician followed the draft ART Bill regulation in India.

Table 11: Distribution of the Participants by Number of Children

S. No	Number of Children	Frequency	Percentage
1.	1	09	27.3
2.	2	16	48.5
3.	3	05	15.2
4.	4	03	9.1
	Total	**33**	**100.0**

(Source: Primary data compiled from data collection)

In India, the two children norm has been in practices and majority of surrogate mothers reported two children, though there were also some families found with three children. It is observed that women, having at least two children, showed interest to accept surrogacy, if their children were born under normal delivery. It is informed that as per the ART Bill, surrogate mothers can go up to five children, including children of their own and beyond which they may not be permitted to go for surrogacy. In this study, participants recorded more than three children.

Occupation

Women are treated as secondary citizens in their work and occupation and they are de-valued in the patriarchal system. Women are mostly associated with household and reproductive works though it is changing in the current context. Occupation of the participants was used, to identify the nature of job and whether the participants received sufficient income from their work.

Table 12: Distribution of the Participants by Occupation

S. No	Occupation	Frequency	Percentage
1	Home Makers	18	54.5
2	Cook	01	3.0
3	Construction worker	03	9.1
4	Domestic Helpers	06	18.2
5	Others	05	15.2
	Total	**33**	**100.0**

(Source: Primary data compiled from data collection)

In the distribution of the participants on the basis of occupation, the home makers constituted more than half of the participants. That means, earning was done only by other members. Others were engaged in paid work but the wage or salary drawn, would reveal how supportive it was to the family. Feminization of poverty led women to undergo surrogacy, in which the reproductive technologies also reinforced the patriarchal views of the society. Women, who underwent surrogacy, conformed to stereotyped gender roles and low economic profile jobs. More than 50% of women were home makers i.e. economically unpaid activity and hence they were pushed into reproductive role, as paid activity. Time and again, it is proved that poverty had pushed women to opt for surrogacy.

Annual Income of the Participants' Family

Annual income of the households determined the status, in terms food consumption, education, type of shelter etc. Income is one of the key indicators, to identify the class background of the participants. Annual income of the sample households was analyzed, by taking into account the possession of movable and immovable assets. Income was one of the most critical reasons for the participants' entry into surrogacy. Household income was calculated as the cumulative amount of the average earnings of both wife and husband or other earning members. Given that majority of participants were home makers and did not engage in paid employment, their size of the family was relatively higher than the average and no wonder the annual income will be an important determinant, which must have forced the women to accept surrogacy.

Table13: Distribution of the Participants by House hold Annual Income

S. No	Annual Income (Rs in L)	Frequency	Percentage
1.	50,000 -99,000	15	45.5
2.	1,00,000 -1,49,000	07	21.2
3.	1,50,000 -1,99,000	08	24.2
4.	Above 2,00,000	03	9.1
	Total	**33**	**100.0**

(Source: Primary data compiled from data collection), Average: Rs: 3.03L.

The average income of the sample participants was Rs.3.03lakhs. A little less than fifty percentage of participants earned less than Rs. one lakh per annum, which was a meager amount and others earned less than Rs. two lakhs. Given that the majority of the participants were from urban areas and with more than the average size of the family, it is quite difficult to survive with such minimum income. It may be underlined that the economic crisis pushed women into choosing surrogacy and hence they readily opted for surrogacy.

Household Monthly Expenditure

In any situation, the income earned must be more than the expenditure, in order to avoid deficit. In case of deficit, the households may seek borrowing, which may not be paid in time and hence they go multiple borrowing. Such continuous and multiple borrowings, keep the households economically backward and they remain in such a state often due to considerable spending on non-essential items and increasing inflation of essential items. Household expenditures constitute expenditure on food and non-food items such as medical, clothing, housing, electricity, water, transport, communication, fuel, education, social and religious functions, marriages etc.

Table 14: Distribution of Participants by Household Monthly Expenditure

S. No	Monthly Expenditure (in Rs)	Frequency	Percentage
1	Below 15000	05	15.2
2	15,000 to 20,000	11	33.3
3	21,000 to 25,000	05	15.2
4	26,000 to 30,000	04	12.1
5	Above 30,000	08	24.2
	Total	**33**	**100.0**

(Source: Primary data compiled from data collection), Average: Rs: 22,250/-

The average expenditure of the sample monthly household was Rs: 22,250/- Given that the average expenditure had been more than Rs. twenty thousand and certain families reported expenditure of more than thirty thousands, there was bound to be deficit every month. To make up the deficit sample households resorted to borrowing. It is found that as high as 90 percent of the households' income was spent on food and the rest on non-food items. Among the non-food items, major proportion was allotted to children's education as well as social and religious functions, including marriages. Such conspicuous consumption accounted for the monthly deficit. In other words, such poor income of the households forced them to opt for surrogacy.

Monthly Expenditure on Alcohol consumption by the Partners

Majority of poor households remained in poverty because income earned by the labour class, particularly men, went to unproductive expenses such as smoking, alcohol, pan consumption etc. Hence it was always a challenge for such households, particularly women, to struggle to have both ends meet and they were forced to take up some kind of work for their livelihood. Construction work, etc, were taken up by women in this study.

Table 15: Distribution of Monthly Alcohol Expenditure by Participants' Spouse

S. No	Monthly Alcohol Expenditure (in Rs)	Frequency	Percentage
1.	Non Alcohol Abuser	10	30.3
2.	1000 -2000	05	15.1
3.	2001-4000	05	15.2
4.	4001-6000	07	21.2
5.	6001-8000	04	12.1
	Total	**33**	**100**

(Source: Primary data compiled from data collection) Average: Rs: 3439.39/-

It is quite surprising to see that the monthly spending, on alcohol ranged from Rs. 1000 to Rs. 8000 maximum, which clearly indicated that object poverty in which the sample households lived, was primarily due to such unproductive expenses, which cannot be justified by any reason. The alcohol drinkers never realize that it puts burden on the household and women and children suffer due to poor home taken income or no income. Such situation often forces women and children, to take up some paid employment. Given that women do not gain any skills in their lives, they just take up occupations, which do not really fetch them attractive income nor dignity of labour, like house maid job or construction work etc. It was observed that more than two thirds of the participants' husbands consumed alcohol, on a regular basis (every day or nearly three or four times a week). The highest amount of Rs. 8000 was observed and depending on the economic status, the brands of alcohol consumed varied to maintain their social status. The most common tragedy, in the families of the alcoholic husbands, was the "pay-day" drinking. This involved a pattern of intense drinking on the day that they received their wages. Whether the person was a daily wager, weekly salaried or monthly salaried, once the wage or salary was received, the liquid cash, available on the day, was spent on purchasing alcoholic beverages, that resulted in deficit in the home budget. As already revealed, consumption of alcohol diverted the income available for necessary to non-necessary items. In the extreme cases of becoming addict to such behavior, it may drain all the income of the household, to go for drinking alone. Thus households were put to financial hardships, as found in the present study. It was found that nearly half of the participants' households remained in financial hardships and they were not able to meet even the basic needs like food. Hunger became the rule and it resulted not due to non availability of job but due to loss of income diverted to alcohol. The alcohol abusers did not take the responsibility of managing the household and instead the wives of alcohol users had to face lots of challenges to provide for the family. It ultimately pushed the women to multiple burdens and responsibilities.

The situation of scarcity triggered by alcoholic husbands, forced the participants families to pledge their properties or valuable item, which was reported by more than one fifth of the participants in this study. In some cases, they borrowed money at high interest rates. To repay this money and fulfill the basic needs of the family, women were indirectly forced to practice surrogacy for commercial purposes. Intermediates could convince more number of participants because of their vulnerability to become surrogate for the well being of the intended parents, agents and doctors. In the surrogacy process, the beneficiaries were intended parents, agents and doctors rather than the participants.

It is inferred that expenditure pattern of the participants' family included consumption of alcohol by the husbands. This played a vital role in the selection of surrogacy as an economic source, for the survival of the family. Liquor consumed by men are socially accepted. The liquor-consuming habits of men, left women with double burden. No wonder women opted for surrogacy. However, it cannot be justified that the loss of income from male members forced women to accept surrogacy because women might not have been willing to accept but for the sake of paying back the accumulated debt.

Savings and Debt

The expenditure and income of the households often did not match and they ended up in deficit. Hence economically backward households often borrowed even to meet daily expenses. In India, the household savings considerably contributed to the national savings. Hence the data on savings and debt position of the sample households.

Table16: Distribution of Participants by Savings and Debt

S .No	Savings	Frequency	Percentage
1	Yes	02	6.1
2	No	31	93.9
	Total	**33**	**100.0**

S. No	Debt	No. of Participants	Percentage
1	Yes	26	78.8
2	No	07	21.2
	Total	**33**	**100.0**

(Source: Primary data compiled from data collection)

Table16 reveals the distribution of participants by Savings and Debt. As far as the borrowing status was concerned, it was found that except one fifth of the households, almost all others reported debt but the savings was almost non-existent, except among two participants. In general, the poverty ridden households resorted to borrowing from the money lenders, private finance, pawn brokers etc, who charged exorbitant interest, which confined such households ever remain in the vicious circle of debt. Borrowing to pay off an outstanding debt with huge interest rate, never relieved them as they would have other expenses to meet. If there was any emergency expense, the second borrowing was made, which added to the economic burden. When the first loan was not paid in time, they may be treated brutally and hence a third loan was sought to pay off the first loan. This was a vicious cycle where the poor households got trapped with continuous borrowing. The fact is that major portion of the earnings went to interest payment. This was coupled with men spending on alcohol and failed to bring their wage to meet their basic needs. As a result, it was quite possible over the time that the borrowings led to a huge cumulative amount, along with the interest. Under such circumstances, called the Debt Trap, women would agree to surrogacy, which they would otherwise not have opted. Self Help Group has been a source of support to these women but, only micro loan could be taken and hence the poor household's reliance on money lenders, for huge expenses, could not be done away with. It may not be an exaggeration to say that indebtedness played a vital role in the surrogate decision. It is also heartening to note that a small percentage of the sample households (22%) did not have any borrowing, but still they would have had other requirements or their income would be just the bare minimum. Whatever the nature of borrowing and the status of sample households, the agents capitalized such vulnerability of the sample households, to draw the women to surrogate market.

Section II: - Process of Entry to Surrogate Market

Having presented the socio economic profile of the sample women, the following section examines the process through which the women entered into surrogacy. It tries to explain the process, right from obtaining the source of information, the process of self convincing, and convincing others, the motivation and negotiations made, the type of coping strategies chosen, etc., Further an analysis of the gender perception and awareness has also been presented.

Source of Information

In this part, the Researcher proposes to understand the source of knowledge, accessed by the surrogate, to enter into the process of surrogacy. Surrogates, from various Districts of Tamil Nadu, had heard about surrogacy from news paper advertisements and some other sources such as mediators, associations, nurses, various personnel working in surrogacy clinic, surrogates as agents, egg donors, physicians, family members of surrogate, relatives, friends, surrogacy related TV serials and films, bit notice circulated by associations and web sources. Majority of surrogates came to know about surrogacy from the neighborhood, or the donors of eggs and brokers or mediators.

Once the advertisement is done, the mediators directly, contacted the women, who were economically weak and vulnerable due to borrowing, lack of income to maintain their family, emergency needs to be met, etc. They were brought to the hospital for egg donation. A few months later, the mediators, contact such women as they have got already informed about the desperate economic needs of these unfortunate women, persuaded them but to help a childless couple to have a child and the huge payment would neutralize their heavy indebtedness. In such a situation, the women were further convinced by the peers, who also indirectly served as agents in the process and received commission from the mediators. Though in the beginning, the women were not willing and a bit reluctant due to cultural questions, 10 months of absence from home, friends, neighbors, relatives etc and also the health aspects of the surrogate etc. But the economic pressure really prompted them convince to accept the offer as that kind of money cannot be earned in their life time. Moreover, such huge money could improve the educational status of children and above all to meet the livelihood needs when husbands' income was not adequate for their households. In the words of a surrogate mother SM9, "*In our neighborhood area, there is an old woman Shanthi who introduced our area women about the egg donation and surrogacy. She had connections with the mediators and agents and so connected surrogate women to such agents.*"

SM8 said, "*I have red about surrogacy from the Tamil daily news paper. There was an advertisement, given by the intended parents where the intended parents also mentioned about their preference for a particular caste to serve as surrogate and provided with their contact. At that time, my family was struggling economically so immediately I contacted them. After the phone conversation, they gave me their residential address for further discussion and process began and I agreed to the terms and conditions and served as surrogate*"

SM12 stated that, "*First I heard about egg donor from a woman broker in our region. After the egg donation, the Doctor told me that my uterus is good, will be suitable to undergo surrogacy? And the Doctor explained about entire surrogacy process and the kind of payment will be made. I agreed and committed to serve as surrogate mother.*"

From the anecdotes of the surrogate women, it is clear that in addition to the agents, mediators etc, the commissioning parent's also directly advertised News paper or TV about the caste preference is also expressed and the surrogacy process begins. From the data, gathered through interviews, it is also observed that not much persuasion had to be exerted because economic pressures made the women to accept the offer.

Television plays a vital role in serving as a good source of information on surrogacy. Every month, associations advertise in the news paper, covering some particular area. Associations and clinics keep giving advertisement for egg donation and surrogacy. Associations organize mediators and volunteers to circulate the printed brochures notices on surrogacy, to individual households in the region where the agents and brokers feel that there will be response positive. Notices will have attractive information, simple process, lump sum money, no side effect, helping a childless couple with child, etc which all will serve as convincing factors.

Image 1: Daily Tamil News paper advertisement for Egg donation and Surrogate Mother

In that notice, they valorize surrogacy, "Thaanathil siranthathu annathaanam, athaninum siranthathu raththathaanam athaninum siranthathu karumuttaithaanam. Eppadi yendral annathaanam seithaal antha neyrathil pasiyai mattum pokkum. Rathathaanam oru uyirai mattumey kakkum. Aanal, karumuttai thaanam seithal oru kudumpathukku pillaivaram koduthu antha kudumpathiayae kaappatriyathu poolakum." (Such matters in the notice will certainly provoke any normal being. It goes like this: "Food donation to quench the hunger is the best of any donation which would address one time hunger, still a better donation would be blood donation which would save a life and even better is the egg donation would gift a family with a child and serve as a future protection for the family").

Based on the above emotional dialogue, as the very first step, the women mediators or volunteers convinced women for egg donation. During egg donation, the clinics assessed the condition of the uterus status of women. A few months later, women, whose uterus happened to be in good condition, were asked to undergo surrogacy by the doctors.

There are several factors which played an important role in the surrogate's decision-making process, in terms of entering a surrogacy arrangement, prior to accessing information from surrogacy centers and doctors, the payment process, the legal binding, practical concerns, etc, which would be second step in the process.

Factors influencing Surrogacy

There are several factors, which act as both facilitative factors and constraints, become surrogate mothers. As already mentioned, it included social, economic, cultural, political, humanistic and psychological factors. But more than everything, it was the economic factor, which motivated or pressurized women, to accept surrogacy.

a. Initial Attitude towards Surrogacy

It is worth examining how the surrogates felt when they very first informed about surrogacy. Majority of surrogates thought that relinquishing the child, in favour of for the childless couple, is a service, though the process is hard. Even if it is taken in the sense of serving the childless couple by the surrogate mothers, the society, the family members, relatives, neighbors, friends, etc may not accept that it is to serve but only to earn money. It can be understood better from the very voices of the Surrogate mother SMI.

SM1 said, *basically I am a lab technician and I have very much interested in Science and Technology. When I heard about surrogacy, it gives me positive attitude towards the technology and also I think that that technology is providing future for the infertile couples.*

SM5 said *initially, I did not have any idea about surrogacy and I was neutral (half minded) about surrogacy. Initially, I don't have much awareness regarding surrogacy and due to poverty, I thought before deciding to do surrogacy to know about surrogacy. Later the association's members explained me and convinced me to do.*

SM9 shared that *when I was first heard about surrogacy, I was scared and was skeptic. Also it creates end number of questions in mind such as its practicing surrogacy is correct or not? Whether the society will accept or not? etc., At last, I got more information through the associations and convinced by them.*

It is clear from the voices of the surrogates that majority of surrogate mothers' attitude was skeptical and doubtful when they very first heard about surrogacy. Some of them had neutral and negative perceptions about surrogacy. The surrogates mentioned that before they met the representatives of the Association, working for admitting women to surrogacy, they were mostly half minded but after discussion with association's members, they got a clear idea about the entire process of surrogacy and got convinced about surrogacy.

b. Emotional experience of the surrogate mother pre-pregnancy

Before the start of the surrogacy procedure, different kinds of emotions were experienced by surrogate mothers like fear, joy, anticipation, surprise, anger, trust, sadness, disgust. As discussed earlier, surrogates had decided to undergo surrogacy due to the pressure of poverty and indebtedness but many of them shared that they had feared surrogacy and its procedure but still had to accept on economic considerations.

According SM08, "*Before my pregnancy, I had experienced fear because of the long process and also social consequences. I had in mind several unanswered question including as to how to remain strong till the delivery? How to face the society? etc,.*"

From the sharing, it is again learnt that no surrogate mother just agreed to the proposal. They had lot of queries to be answered to self convince like the fear about physical strength, fear to face the society etc. Somehow, the marketing process rested with the Associations, who, by showing previous examples, explain the simplified process and the service dimension of surrogacy, really motivated and encouraged the women to accept the surrogate proposal. The economic need on the supply side, the market on the demand side, come to an equilibrium, where several non economic factors do influence the decision making process.

A few of the surrogates shared that before pregnancy, they were happy as the surrogates had a feeling that the childless couples will be blessed with the child, which is a heavenly service indeed. In certain cases, the surrogate mothers went to the extent of saying that if we had adequate economic support in the family, the surrogacy would have been undergone free of cost, to literally support the infertile couple, to have a child. Feminist argue that women are still kind hearted, though their economic conditions drove them to be market oriented. It is very difficult to bring women into the marketing process because women were always associated with nature.

c. Family forcing the women to serve as Surrogate

Surrogate is normally persuaded to undergo surrogacy, by mediators or by their family members, to promote their interests. To check whether the family members forced the surrogate to continue the surrogacy as a commercial activity, the Researcher asked the participants to share their experiences. It was found that no one was forced by their family members. Even husbands did not favour surrogate practice because a few of them considered the health of the surrogate and a few other perceived surrogacy as an illegal work. No one, however, thought that it amounted to prostitution as other men's sperm is given to the embryo, which is being borne by their wives. Or even if they thought so, it was not expressed due to economic considerations. Poverty and hunger, food insecurity, starvation of the children etc, were the immediate need to meet their livelihood and the cultural impositions had to take the back seat.

As it was held by SM01, "*In the beginning, I tried to get permission from my husband because our marriage is self arranged, so it was not issue to wait for other members of the family to consider such proposal. However my husband did not give permission immediately to take up and agree to surrogacy because he thought if I undergo surrogacy, my health will be affected but later on, I convinced him to undergo surrogacy explaining that it won't imply on my health.*"

According to SM027, "*Once I came to know about surrogacy I conveyed my wish to my parents and family members but parents immediately refused to permit surrogacy. Also, they said your children need their mother, if you practice surrogacy your children will suffer. But, I took them into the association and they took few days to convince my parents and also my children.*"

It is worth reiterating that lot of convincing had to be made, where none of the family members had allowed for surrogacy initially. Later, family members were convinced by women who opted for surrogacy. Associations also informed them that serving as surrogate would not affect the health but only will bring mutual benefit, to address the respective of needs of surrogate and intended parents. Many surrogates practiced surrogacy for the welfare of their children and their family and not for their own development. It seems that women were always concerned with the well being of the family. Patriarchy makes them feel guilty when they do not come to the rescue of the family distress and hence women always take it as their duty, to help by all means the family and safeguard the interests of the family despite against their wish or health etc.

d. Factors associated with decision to serve as surrogate

Surrogacy does have cultural conflicts, particularly among relatives, family members etc. Many women take life changing decisions but taking such a decision on surrogacy was quite difficult and challenging. There are some motivating factors such as livelihood, poverty, unemployment, debt, education for children etc. which triggered such decisions on surrogacy. Women do not mind the social pressures because they prioritize the economic needs. In this study, the surrogate and her family were engaged in low paying, seasonal and informal work. Surrogate did not get basic means of livelihood and faced financial difficulties. Under these circumstances, the respondent surrogate had to borrow money at usurious interest. The remuneration for surrogacy, nature of payment (lump amount) and the short period for which the amount was received, were major reasons for accepting surrogacy.

SM9 said that, *I was divorced and my two girl children are living with my parents. Now my parents and brothers were taking care of us but how long my parents and brothers will take care economically. If my parents get older and my brother gets married, it is no longer possible to get similar support. What will happen to me and my children? Against this background, I thought I would become a surrogate. It is money that gets you to do everything. She also highlighted the lack of support in raising children and securing their future. Hence easily accepted and opted for surrogacy.*

SM5 mentioned that "*Because of my family hardship, I became a surrogate. Also, my husband was addicted to alcoholism and to fulfill the basic needs such as house rent, I had to undergo surrogacy. Also, my family faced problems due to financial difficulties.*"

SM33 shared that "*due to alcoholism, I lost my husband four years ago. I have worked as a cook. We are four in the family and my income is not sufficient to run the family. In this situation, I would like to act as a surrogate for our livelihood and to educate my children.*"

With informal seasonal jobs, neither the surrogate nor her husband could get such a big amount, within a short period. But surrogacy helped them to get a lump amount, within a short duration. To earn such an amount, surrogate and her husband might have to spend a number of years. It is not possible to get such a huge amount as a single payment. But in a single surrogacy agreement, they could earn such a huge amount. Getting such a huge amount was a crucial part of the surrogacy decision, taken by the surrogates. Few of the surrogates had received remuneration, that was lower than her family's annual household income, for the corresponding period because those surrogates were cheated by the brokers. SM16 said *that her husband was addicted to alcoholism and for his continuous drinking purpose, he needed money every day. As a result, he sold our house. The money which, she got as remuneration was used to purchase back the house which was sold by her husband.* The most important thing was the surrogate being burdened with loan for high interest. Poverty and debt indirectly forced women to become surrogates.

e. Reason to become surrogate

There are various reasons to become a surrogate. It should be mentioned that surrogacy is looked upon as a stigma and a taboo. Hence there may be various reasons to undergo surrogacy but taking such a kind of decision was very painful. The reasons were to help childless couples, opportunity knocked at the door, attractive money offer, offer of huge money, improve the status by taking surrogacy as profession etc,.

SM1 said, "*I was very much interested in science and technology. We have a child and thus we had completed our family. But my husband wanted to see me as a pregnant woman but we were not interested in the second child. At this point, we decided to undergo surrogacy so that we can fulfil our desire and also the childless couple also got their child. In this situation, opportunity knocked at the door.*"

Majority of surrogates opted for surrogacy due to attractive money offer as most of them were in huge indebtedness. Some of them needed money for medical and school expenditure of their children, which was the bare minimum and no parent would want to impose their poverty on their children's education.

SM9 said, "*My son had a problem in heart and to save my son, Doctor advised us to do a surgery. For that I was desperate to meet up the monetary requirement for the surgery. Under these circumstances I had no other go to earn, to save my beloved son's life. Hence, I immediately offered to the offer to surrogate mother proposal as I felt that only buy mobilizing adequate money, I could spend on the heart surgery which would be possible only when I agree to the proposals like surrogacy.*"

If she had been economically better off, she would have done it as service, she added. But due to pressure to earn money for son's heart surgery, she had asked for the payment. '*Only a woman knows the pain of another woman and we would have given the happiness to another family*' she added. But the fact is that surrogacy does help in such vulnerable situation, which has mutually helped in giving life to a foetus and saving life of a child suffering with heart issue.

f. Method of self and external convincing

The method of self convincing is most important to undergo surrogacy. Procreating for a third person and caring for another woman's child, is a service but still it may not be agreed easily by everyone, particularly due to the taboos associated with the surrogacy.

SM9 shared her experience as follows, "*Due to huge debt I have convinced myself to undergo surrogacy. A few years back, I lost my son and I have only daughter. So I thought, at the end of surrogacy, if we get 50,000, that's enough. I planned to invest that amount for the secured future of my daughter.*"

SM6 in her experience held as follows, "*Threat of huge loan and attractive money were the major reasons to convince myself to procreate and hire out my uterus. When agents offered huge money, due to poverty, I accepted to undergo.*"

Women, who acted as surrogates categorically, held that usually women are very soft in nature and they will show empathy to another woman who is suffering due to her inability to procreate. As a surrogate, we are benefitting them and at the same time, they are also benefitting us. Securing the future of their children was a common aspiration, expressed often by the surrogates during interviews. The surrogates insisted that they were undertaking surrogacy for the sake of their children. It is not a surprising in the Indian culture that the children are the priority and even after suffering with the man, the woman still continues to live with the same man for the sake of children.

Birthing Agreement

a. Nature of contact between surrogate mother and clinic

Legally speaking, surrogate mothers must disconnect and they are not permitted to have any relationship with intended parents, post surrogacy. However, there are kind hearted parents, who do not mind the surrogate to come and see the child whenever interested. It is interesting to examine how surrogate mothers are treated in the beginning, middle and post surrogacy. Agents gain commission, doctors get cases and the surrogate mothers get the money, to meet the livelihood or emergency needs. This trio actually builds the relationship. In majority of cases, the post surrogacy care is minimum or nil. In this study, there were mixed kind of responses. In the case of foreign couples, it may not be possible for the surrogate mothers, to see the child in person. In this study, it was observed that the role of the clinic and the infertility physician was vital in the case of surrogacy. They act as the pillar of strength to the intended parents, mediators and surrogate mothers. In majority of cases, surrogate mothers were either approached by the agents or they had opted for surrogacy through advertisement in the media or some of them had heard from their friends or from the clinic during the egg donation. As discussed earlier, the agents had arranged few volunteers to approach women, who were economically vulnerable, from the surrogacy perspective.

SM5 said, "*I was not directly contacted by the clinic. The agent asked me to go to some private hospital and as per their guidance; myself and my husband went there. When we reached the hospital entrance, I was surprised because the hospital was too big and I did not enter that big hospital. Also, I wondered, whether they will treat in a good manner? But, after entering the hospital, the staff treated me decently.*"

Majority of surrogate mothers found the hospital staff to be cordial when they first approached. From the agents and doctors perspective, the surrogate mothers were merely like "money banks" for infertility doctors and agents as they reaped huge returns. Although they mostly cared for the well-being of the baby, the emotional care of the surrogate mothers did not hold any importance to the infertility physician. Hence the surrogate mothers remained deprived of the emotional support, which is normally given by her family, during her own pregnancy.

This is the huge gap between normal pregnancy and surrogacy pregnancy. In the normal walk of life, the pregnancy is cared for with focus on concerns of health of both baby and mother. In the case of surrogacy, the health of the mother is approached with the commercial perspective of safe delivery to get the child. All the more challenging was the post surrogacy because there was emotional need during pregnancy as the surrogate mothers were far away from their family, children. Hence surrogates were not comfortable as they regretted for having accepted the surrogate proposal, though economic need or absence of earning members etc compelled them to do so. So it is necessary that the surrogate mothers are given all comforts, due to the concern of both the baby's health and the mothers too. It is necessary that some kind of support system is inbuilt into the payment for post surrogacy needs.

b. Place of Stay- During Pregnancy

Due to the cultural and ethical issues, surrogates did not like to disclose their identity as surrogates and the pregnancy had to be explained to the people with whom they interacted on a daily basis. This often compelled them to offer explanations for the pregnancy as well as the giving away of the child, or to make themselves physically unavailable to conceal the pregnancy. Yet it was apparent that surrogates negotiated with their circumstances in different ways, assessing risks in various ways. Otherwise surrogates were excluded from the society because society thought that surrogacy was a dirty work. All these factors determined that nature of stay was very important from the above mentioned points and also on the surrogates' mental ability during surrogacy because majority of them concealed their role, from the relatives, due to the social stigma. Many surrogates stayed at home, arranged by the agents near the clinic. Some of them stayed in their own house, with their husbands. Some of them stayed with their family in a temporary tent house. A few of them were asked to stay, with the intended parents and their relatives. In other words, the participants revealed that there was no compulsion as to where the surrogate should stay as this study revealed that the surrogate mothers had the freedom to stay with their family, parents and with husbands. Only requirement was that proper care must be taken to deliver the baby. The stigmatized environment made them to stay away from home. But, in majority of households it was accepted as normal and it was not an issue. Such emotional comfort and not feeling alienated and by staying from home etc will help the surrogate to deliver a good and healthy baby. Though many studies observed that the surrogates were almost imprisoned in homes till they delivered the baby and strictly monitored (Ponde, 2014), it was surprising to observe in this study that not all the surrogate mothers were strictly confined at home. Surrogates felt at home and did not suffer any psychological illness when they stayed with their family.

The neighbours must have been informed and it did not become an issue and the society did not ridicule them. Whenever there was problem in revealing the identity of surrogate mothers, they preferred to stay away from home.

There must be anxiety among intended parents, to have the child delivered, without any damage. Hence a conducive environment, both physical and mental, was provided. The intended parents provided all comforts to the surrogates till the delivery and kept them in their custody or in their relative's custody, whichever was convenient. One of the surrogate mothers shared her experience, as follows:

SM29 said, "*I stayed with the intended parents because they would like to stay with me for the welfare of their child and feel the growth of the child.*"

SM33 said, "*I was asked to stay with intended parents' relatives, to take care of the child. We stayed in temporary, rented house near the clinic.*"

The stigma that surrogates faced, over the surrogate pregnancy, consequently structured the practice of surrogacy in a way that had a direct bearing on their position in the arrangement. Agents were responsible for arranging the residence, which was often in a location conducive to the supervision of surrogates. The shift to a new place brought the surrogates under the direct control and supervision of agents and mostly surrogates were accommodated near the clinic to attend to their needs, if any, in the case of emergency situation. Meanwhile, to minimize the residential expenditure, the agents rented a big house in which many surrogates were housed. But from intended parents, each and every month, the agent collected a sum of money, separately for each surrogate, towards accommodation and food. The shelter home provided the basic requirements to the surrogate such as cot with bed, fan and television for entertainment and agents provided food. The agents were responsible for the health of surrogates and the child. The mediators considered that every surrogate as a money making source and therefore, each head was counted though the common facilities were shared. The intended parents, who were desperate to have a child, did not make it as an issue but paid just what was actually asked for. There were different kinds of exploitation, with different stakeholders, including doctors, on which the surrogacy market was flourishing.

c. Parties included in the Written Contract

All the participant surrogate mothers mentioned that surrogacy agreement, between all the parties involved, took place in the form of a written contract. It was mandatory to get the signature from the surrogates and the family members such as husbands or parents of surrogates, before entering into the process. The clinics or the intermediates normally prepared the agreement, with the help of an advocate. The contract was in a bond paper and the agreement copy was prepared in the 100 rupee stamp paper. Normally the full agreement was neither fully explained nor read by surrogate mothers but orally informed about such provisions, terms and conditions to be followed. The voices of the surrogate mothers are given below.

SM1 said, "*Each and every page of agreement myself and my husband were asked to sign and it was prepared in a stamp paper. In that same agreement, intended parents, the clinic, advocate and the agents also signed.*"

SM29 said, "Myself and my father signed the contract copy and they asked my mother to sign as witness. The agents also signed the agreement."

The surrogacy process was carried out after the signature was obtained from the surrogates and their family members. But many surrogates did not know who had signed the contract. The surrogate did not know that intended parents, doctor, advocate and intermediates had also signed the contract. The association representative or hospital staff usually obtained the signature from the surrogates. In other words all pages were not read out to the surrogate mothers before their signature and hence there was a gap between written promise and actual performance.

d. Issues addressed in the written contract

The agreement copy normally contained all information, in more than ten pages. In ten pages, various kinds of issues are presented as written contract. But during the interview, the surrogates gave different kinds of answers, which were quite shocking because many surrogates did not even read or remember what those written contract contained. In the words of the some of the participants:

SM3 and SM7 said, "*I don't remember what clauses had been mentioned in the written contract.*"

SM2 and SM6 said, "*Except the compensation, I don't know what are the clauses mentioned in the written contract because intermediates informed everything to my husband. Also, myself and my husband were not able to read the written statement because the written contract was not in the vernacular language. Since my husband could not read, intermediaries just explained and we agreed and singed. So we were not sure of what matters were there in the written contract black and white.*"

Surrogates signed the agreement but many surrogates were not aware of the issues stated in the written contract, except the compensation. Also, the surrogates could not remember the issues mentioned in the contract. Many contracts were prepared in English and few contracts were both in Tamil and English. As it was discussed earlier, majority of surrogates could not read and no wonder they did not understand the medical terms or complicated procedures that might affect the health and well being of the surrogate mother. Moreover, even the clinics or association did not leave a copy of the agreement with the surrogate mother, who is a signing party to the agreement, to have any evidence either of her pregnancy or the surrogacy arrangement. Thus it is not different from any other labour contract, where the matters mentioned in the contract, will never be properly explained nor could to be read by the persons concerned. The surrogacy arrangement is not an exception.

e. Copy of written contract to surrogate

Surrogacy agreement was made between the surrogate and the intended parents. The agreement was prepared by an advocate and the agreement contained a minimum of ten pages. The main part was in a 100 rupee stamp paper while the remaining pages were in normal sheets. In the agreement, different clauses were mentioned by the advocate. All clauses contained medical terms, the meaning of which surrogate and her husband were not aware of. Under these circumstances, surrogate and his husband could not understand the medical terms and the clauses were orally explained by the agents or advocates. It is worth noting that a copy of agreement was not given to the surrogate.

This gap was expressed by almost all the surrogate mothers. "*Myself and husband or the parents only signed in the contract. After that they never gave any copy of contract to us.*" Even though, a few of them held as: "*I have asked the contract copy from association representative but they said that in the contract intended parents' address was mentioned and so I cannot give the agreement. I will keep the contract copy*" As per the norm, the intended parental address should not be given and hence it is given as a reason for not giving a copy of the contract. But if that is the reason, it is always possible to get the address hidden. If there is a will, there is a way. But some lame excuse is given, as mentioned above.

Normally, in all the agreements made between two parties, the contract copy should be given, to both parties, for their future reference. But in surrogacy, surrogate did not get a copy of the agreement. It clearly shows that the entire surrogacy process was done for the welfare of the intended parents and not for the surrogates. Again, it clearly indicates that surrogacy was open to exploitation, to the advantage of the intended parents rather than to the surrogates. The same kind of excuse might be given to indented parents also by telling that the addresses of the surrogates are presented in the agreement and so copy cannot be given. It is understood that the exploitation happens both in terms of fixing higher charge with intended parents and lower payment to the surrogates. To hide such facts, the copy of the contract is not given nor is an opportunity given to read the document. This non- disclosure of contents of the contract, to both the parties, opens the way for rampant exploitation.

Coping Strategies during to stay in Surrogate Hospital/Hostel

The surrogate has to stay away from her own home, for approximately ten months to undergo surrogacy. These women should be under medical observation from beginning to end. For this reasons, the surrogates had to be accommodated in surrogate hostel, which comes under the direct care of the clinic. The clinic does not have such facilities and hence the mediators or the intended parents arranged separate homes near the clinic, for their direct care and continuous check up. To safeguard the child, intended parents arranged sophisticated homes to the surrogates. Once the physician confirmed the pregnancy, the surrogate will not be allowed to go home though in certain cases, the surrogate mothers had the liberty to stay with the family, at home.

SM2 and SM29 said, "*My husband often visited me and for a few days, stayed with me to take care of me. I regularly contacted my parents and the remaining time, I spent time to watch television programmes. I prepared my mind to stay aloof till the delivery away from any troubles or disturbance and not to feel isolated for the sake of safe delivery of the child and hand over to the intended parents.*"

SM1 said, "During surrogacy, myself, my son and my husband stayed together in a separate rented house and the rent was paid by the intended parents."

SM3 and SMS said, "Only in the first month they asked me to stay in the hospital and the remaining months, I stayed in my own house."

Hence coping strategies have been provided by the intended parents and they were flexible enough to keep the surrogate mothers as comfortable as possible by giving them the option to stay with family in the sophisticated rented house or in their own home of surrogates, as per their wish. Such flexibility and liberty might have comforted the surrogates and it would not have been necessary to seek coping mechanism. But this was not true with all surrogate mothers, as some of them were forced to stay in a particular place provided and away from interacting with family etc. In such cases, the coping strategies were a bit difficult.

Surrogates, whose marriages were not arranged by parents, did not have any relationship with their family members and they were neglected by their family members. In this situation, the surrogate mothers were willing to stay with husbands, for support and care. Some surrogates with family, wanted to be away from their family and their children. In such cases, the surrogate mothers had informed their children that they were going to take care of an old woman who needed help and therefore, could not visit them in between and should obey their grandparents till they returned. Surrogates maintained that they were doing all these things for the welfare of their children. These issues of not disclosing the fact, would also certainly affect the mental health and it would be difficult to look for coping up. But women in the name of supporting the family, on the one hand, with economic needs and on the other hand, helping the childless couple with a child, always took health for granted. These surrogates knew how to cope up with. Feminist contention is that patriarchy exploits such characteristics of women. Moreover, the socialization process shapes the gendered human behaviour, where the women are always told to adapt to changing needs of others.

Hence such coping is not surprising in a patriarchal context.

Consequences of Surrogacy

a. Consequences of practicing surrogacy on surrogate family

The practice of commercial surrogacy is shrouded in secrecy and often various persons such as the agents, surrogates and commissioning parents, were hidden from the public eye. Hence surrogate mothers faced different kinds of problems, in family and society. There is surrogacy stigma behind practicing surrogacy. In public life, commercial surrogacy is deemed to be baby- selling and equal to sex work. This stigma compels the surrogate to perform such task, hiding themselves or staying away from their own family and relatives. Due to this misconception, women were not willing to share with others about surrogacy. Globalization has persuaded people to accept invention of science and technology but they are not ready to think rationally, regarding commercial surrogacy. It can be observed, in the voice of the surrogate mothers, in the sample given below.

According SM05, "*If you tell others, people won't let you live. They will abuse saying that she left her home and children and roaming or flying everywhere. It is not good for a family a woman. People may further abuse that she is doing it for money, selling a baby for money. Such abuses may draw to wrong conclusion. Initial stage, nobody knew that I was practicing surrogacy. I didn't inform anyone while I was practicing surrogacy. A few months later, mother in law came to know about my surrogacy from someone and during that time, we tried to avoid them both in person and over phone calls. So they verbally insulted me like I was doing something immoral (sex work), I was not a good character woman and that's why I was avoiding the family members.*"

When things are not disclosed about surrogate practice, one is easily suspected to be doing sex work or abuse to the maximum, putting surrogate mothers into an embarrassing position. However, the husbands normally stand with the surrogate mothers, Majority of the surrogates shared about popular misconception about surrogacy, like surrogate should have sexual relations, combined with the assumption that a child was given in exchange for money and these were the sources of the stigma. The act of bearing children outside marriage and as part of a commercial transaction, went against the dominant perception of reproduction as an activity to be performed within the private, familial space, motivated by love and care, and hence surrogacy is seen as undesirable and 'immoral.'

Similarly, SM 29 said, "*When I informed my parents, they did not agree immediately. In my family circle, everyone knew while I was practicing surrogacy. But, except family members, we did not inform anyone because people entertained misconception about the concept of surrogacy. Also my children were getting restless due to my absence. Because of that, often their health got spoiled. Meanwhile, after surrogacy, they took few a months to resume relationship with me.*"

Majority of surrogates shared that practicing surrogate pregnancy was not accepted by the society because surrogacy collapsed the cultural norms of sexual behaviour and reproduction, as constructed and followed within the patriarchal institutions. Confining women's sexual and reproductive capacities to marriage and family, the exchange of currency under surrogacy, led to the labelling of surrogacy as baby-selling and sex work, both of which were also stigmatized on the same basis. Orthodox families considered surrogacy pregnancy as immoral, within the family sphere. But after the globalization and liberalization, women were indirectly forced to cross the boundaries, in the reproductive market, as objects which may be permissible in the global context. The culture has no longer an issue for the surrogate mothers because confined to as they are desperate to earn money for their family and at the same time helping a childless couple.

b. Decision regarding abnormality found in foetus

If any abnormality is discovered through ultra sound scanning, immediately the foetus is aborted. Such cases may happen in surrogacy also. This question was posed to the surrogate. None of the surrogates was aware of this particular issue. They mentioned that the doctor and the intended parents would decide. But in the agreement, it is mentioned that some genetic and congenital abnormalities, may not be detected, by the amniocentesis or other testing. It is the intent and understanding of the parties that the surrogate shall undergo a therapeutic abortion or selective reduction, under such circumstances. The intermediates claimed that if the pregnancy were to be aborted or terminated, due to some reasons, as per the agreement, the surrogate mother was often paid half the amount of what she was supposed to get under normal circumstances. It may be a compensation for the kind of treatment undergone but at the same time, it is quite disappointing for both the sides.

c. Reaction/consequences- if abnormality is found after the birth of the child

In some cases, when the delivery abnormality is found, the intended parents can refuse to accept the child. On that basis, the question was raised but answers given by surrogates were hypothetical. It is unethical to leave the child with surrogates if the child is defective. The agents may need to mediate and set right such issues, which rarely occur.

SM1 and SM4 said, "Whatever *be the genetic risk, we are just a carrier, so the intended parents should accept the child.*"

SM7 and SM6 said, "*If the intended parents refused to accept the abnormal child, we will take the child.*"

Though the surrogate mothers were magnanimous enough to accept the defective child, it will certainly be a liability given the poverty, food insecurity, unemployment experienced by the family of the surrogate.

According to majority of surrogate mothers, the commissioning parents would still accept the child, in spite of the abnormities of the child. Although this expression was completely hypothetical, yet it reflected the thought and views of surrogate mothers, who realized the desperation of the commissioning parents. Some surrogate mothers reported that they would accept the child if the commissioning parents or the clinics or doctors refused to take care of the child, irrespective of the fact that they had their own children and poverty to fight with. This reflected the emotional bond between the foetus and the surrogate mother. It was the surrogate, who had borne the child and hence that bonding would certainly be there.

d. Mental Health Consequences

Relinquishing the child for money, may directly affect surrogate's mental health. It erodes women's self confidence and leads to problems such as crying, frightened of noises, depression, suicidal tendency, sleeping as well as eating disorders, anxieties, mental restlessness, etc. Though surrogate may not have genetic relationship with the child, 25 weeks of surrogacy pregnancy develops bond between the child and the surrogate mother. It leads to mental consequences, after surrendering the child. There have been cases reported in this study, where the surrogate mothers developed several mental health issues (Van den Akker, 2006).

(i). Depression

Depression not only affects person's personal life but also affects one's professional life. Depression is a mental health problem, which is estimated to become the second most common cause of disability, after heart disease, by 2020 (Samaritans Information Sheet). The symptoms of depression include psychological symptoms like depressed mood, loss of interest or pleasure in activities, including sex, feelings of guilt, hopelessness and worthlessness, low self esteem and physical symptoms via sleep disorders (sleeping more or sleeping less), appetite or weight changes, difficult to concentrate, lack of energy and fatigue.

In this study, many surrogates faced moderate levels of depression, after relinquishing the child. Nobody consulted a psychiatrist for their recovery and they lived with their mental stress and pain. The depression not only affects the surrogate but also affects the children, husband and relatives and surrogates could not live happily in future.

Taking such issues into consideration where the surrogate families were entirely disturbed, there must be some post surrogacy care to be provided or a payment to spend on post surrogacy symptoms. The doctors and the intended parents and other stakeholders, who are engaged in the process of surrogacy, should not only look into proper delivery of the child but also concentrate on counselling the surrogate, to be prepared to bear the severance with the bonding of the child, as agreed. The emotional bonding, if any developed would only affect their health and even the money allocated may not help if depression turns to severity. For the sake of supporting the economic needs of the households, the surrogate mothers hire our their services and if it turns out to be depression or other forms of illness, the purpose gets defeated where the surrogate mothers suffer with mental illness post surrogacy. Hence, a detachment must be cultivated by the surrogate and professional counsellor service must be available post surrogacy, which must be part of the intended parental budget.

(ii). Sobbing

Humans are the only creatures who are endowed with feelings. It is very clear that strong emotions trigger women to cry. But crying is more than a symptom of sadness. It is triggered by a range of feelings, from empathy and surprise to anger and grief. Based on the relationship bond between two people, crying will differ. Crying can give relief, from aggressive feelings like rage, anger and disgust.

In this study many surrogates reported that they cried to some extent and a few of the surrogates shared that two or three days, we literally kept weeping till the bonding gets slowly gone away. Later we were forced to concentrate on our own children and family.

After relinquishing the child, surrogates might have developed strong bond with the surrogate baby and no wonder they were sobbing to some extent. The most interesting thing is that the surrogate, who had relinquished child two and half years ago sobbed during the interview. It clearly indicated that some experiences of surrogate towards the surrogacy child, do not go away even after the passage of time. Even the strong-willed surrogate recorded her strong feelings towards surrogate child, through sobbing

Experiences gained during Pregnancy

Surrogate mothers experienced lots of problems before and after pregnancy and their different kinds of experience indicated how surrogate mother felt, good or bad. The experience depended on various factors like their relationship with the commissioning parents, place of accommodation, relationship with their husbands, before and during pregnancy, management of their home and children in their absence, emotions felt during pregnancy for the baby, etc. This part deals with such critical issues and examines the pros and cons of surrogacy contract.

a. Frequency of Contact between Surrogate Mother and Commissioning Parents

According to majority of the surrogate mothers, as per the agreement, there should be no relationship with the commissioning couple, before and after the pregnancy. Few surrogate mothers reported that surrogate mothers, who stayed in the custody of commissioning parents or their relatives, experienced terms verbal abuse, humiliation, exploited their labour, by compelling them to undertake all the household duties etc.,

SM33 said, "*Intended parents asked me to stay with their brother and his wife. She prepared food every day and I did not like to eat that food and due to severe vomiting, I just avoided the food but they forced me to eat. Also, because of the injection and tablets, often I had body pain. At that time also, they forced me to sleep whole day but due to severe body pain, I was not able to sleep. Meanwhile, they restricted my contact with my family members by phone. They forced me to do all the things to get their child healthy. They tortured me to the extreme level and also they teased me verbally. Also, I was not able to share all these things with my family members because they did not know that I acted as a surrogate. I kept all these things with myself and it was such a painful situation.*"

In other words, surrogate mothers did not experience the kind of comfort they wished to. The concern of the intended parents was to get the child delivered properly. Towards that purpose, the surrogate must co-operate despite her severe illness, discomfort, vomit etc. It is all the more challenging in such a situation because the surrogate was prevented from contacting their family members. The pregnant mother, when she was not healthy, and not emotionally comfortable, it would certainly affect the health of the baby too. But it was not understood. Moreover, it is necessary that the intended parent should keep the surrogate mothers in their custody instead of their relatives, as the same kind of empathy may not be expected from relatives as the baby was for the commissioning parents. The ill treatment was serious and due to their vulnerability and helplessness, they could not share such agony or pain with anyone but to suffer in silence. Such pressures may lead to depression and mental health post surrogacy. Associations, which looked after these rights of the surrogate mothers, must clearly spell out how the surrogate mothers must be treated and their comfort must be maintained. Otherwise, the surrogate will just be treated as an object and a product to be used and thrown. It is against human right and moreover, having rendered such a noble service, whatever money for which the surrogate was engaged, it is inhuman to ill treat them and cause mental health issues. Surrogate mothers must also raise their voice collectively against such acts of cruelty.

SM29 said, "*I stayed with intended parents from start to end. During pregnancy due to severe vomiting, I was not able to take food but mother of the child threatened me. They warned me that if anything happened to my child they will not spare me. But I explained that I am helpless. My physical condition did not allow me to take food. Meanwhile, she said, your health is important but my child should be healthy. Sometimes, she abused me verbally.*"

The intended parental concern would be naturally with the health of the child. At the same time, these parents who had spent huge amount, to get their own biological child, should also ensure that if the surrogate is not able to physically feel comfortable and eat food, it must be treated rather than and verbally abusing them which would only worsen the mental health of the surrogate as well as that of child.

This downturn in relationship could be attributed to the stress, created for the surrogate mother, who lived in shelter home away from her children and family. Intended parents expressed their concerns regarding the well being of the foetus, which further put pressure on the surrogate mother. Due to these factors, surrogate mothers get depressed during the pregnancy. Surrogacy pregnancy should be considered as high-risk emotional experience because many surrogate mothers faced negative experiences (Tashi, S, et al, 2014).

b. Development of Child During Pregnancy

The Researcher wanted to know whether the surrogate felt growth and movement of the child during surrogacy pregnancy. To know the experience of surrogate, data were gathered on their experience. Majority of surrogates shared that everyone felt the growth and movements of the child. A few of them recorded that they did not mind it because if they felt those things, then they could not leave the child after delivery, even if it was not genetically related to them. Some of them mentioned that before initiating the surrogacy process, they completely fixed their mind and they were mentally prepared to depart the child, which did not belong to them and they did not have any kind of relationship with the child because they were just carriers.

SM05 said, "*During surrogacy pregnancy I felt the growth and movement of the child. Most of the time, the child was too active and at that time, I used to speak to the child like you are playing with me but you are not my child, you will leave me.*"

SM27 said, "*Through pregnancy, the growth and movement of the child was too high. The child played like my child, during sleep time I felt the movement of the child often. In the 7th and 8th months, the child was more active and during mid night in sleep, child would kick on top of the stomach.*"

It is a good emotional experience but having agreed to be a surrogate, it was just a contract. It was really painful to disconnect from the child, which lived with one's uterus. But the commerce does not permit such emotions and surrogate is compelled to remain without any heart. It was learnt that majority of surrogates felt the growth and movement of the child. Some surrogates spoke to the child. This emotional movement and dialogue created a bond between the child and the surrogate mother. This kind of attachment, mentally and psychologically affected surrogate's health post surrogacy.

c. The Worst Part of being a Surrogate Mother

In this section, the pain experienced by surrogates is examined. During pregnancy, some surrogates were emotionally attached to the child and it was a tough job for the doctors and agents, to hand over the baby to the intended parents. Also, due to pregnancy, the children of surrogates distanced themselves from their mothers because of the long absence. The health of the surrogate was also affected after the delivery.

According to SM33, "*The entire surrogacy pregnancy was the worst part in my life because during pregnancy, I was verbally often abused and ill-treated by the relatives of intended parents. I was cheated by intended parents because I had delivered a child six months before but not yet received the money. Because of surrogacy pregnancy, I lost my happiness with family members, my children and my health. Before surrogacy, I was very active but after surrogacy, I could not be active and still I have pain in my whole body. When I was thinking about my surrogacy days, it was too hectic thing to me to become normal. For relaxation, I could not able to share these things with my family members because of the secrecy involved in the surrogacy pregnancy.*"

According to SM01, "*During surrogacy, we (surrogate and her husband) avoided to contact with our family members. Also, to stop internal bleeding, doctors put oil injection every day and it caused unbearable pain. During pregnancy, often I was vomiting. When I informed doctors that I could not eat because of vomiting, they immediately put me on trips. It was also painful. The secrecy involved in surrogacy pregnancy and injection pain etc were the worst part in surrogacy process.*"

Majority of surrogates reported that relinquishing the child after the delivery was the worst part of the surrogacy. Some of the surrogates registered that they wanted to see the face of the child, after the delivery.

SM05 said, "*I expected that after the delivery, the intended parents would convey how I felt after getting the child. I would like to know the growth of the child in future. At least these things the intended parents could have shared through association representatives. If intended parents had done these things, I would have felt very happy and proud of myself.*"

Majority of surrogates expected these kinds of sharing with the intended parents. Surrogates were emotionally attached to the child. SM24 *reported that the foremost important thing was that she delivered the baby thirty months (two and half year) ago but still she was thinking about the child and she became speechless.* All these observations clearly illustrate that relinquishing the child was not an easy task to a surrogate and relinquishing the child, though not genetically related to the surrogate, was painful for the surrogate because she remained a mother. Motherhood made women very weak and created emotional bond with the surrogate child, though not genetically related to her.

Though the study expressed a mixed kind of feelings and emotions with the surrogate mothers, where one section prepared their mind from the beginning while another section really developed a bond and suffered post surrogacy. The unforgivable issue was the abuse, ill treatment and torture being reported during surrogacy pregnancy. No price can be paid to the pain that the surrogate mothers undergo, in the form of injections every day, vomiting and other physical pains coupled with the mental and psychological pain which they suffered with patience only because of the surrogate payment. Market decides that baby can also be bought with a price, but one cannot account for the kind of losses that the surrogate mothers experience and the happiness and peace they had to forgo, which might get reflected in their future life also. These emotional experiences, a fall out of surrogacy was immeasurable and cannot be compensated for.

d. Routines – Missed or Sacrificed

During surrogacy, majority of surrogates were not allowed to follow their routine works. Surrogates were away from home, family members and especially from their children. It is important to know what surrogates sacrificed, to give pleasure for unknown persons and for the sake of money. Incidentally, where sacrificing is concerned, nobody can match the mother. In this context, surrogates were asked to share their experiences.

SM 29 shared, "*During surrogacy, every day twice, I spoke to my children but I missed my children in the eighth month of surrogacy pregnancy. The eight month distance created loneliness and it affected me emotionally and mentally. Because my children always slept with me and I could not sleep without their warmth. Most of the time I missed my children's proximity and when I thought of them often, I suffered in silence and ventilated thought my breaking and weeping.*"

SM05 said, "*Before surrogacy, I was very active but during surrogacy, physician advised me to take complete rest which I hate as it was against my will and nature. During surrogacy, I missed the household activities and preparing food for my family. For the cooking process, I used to go out to purchase. But because of pregnancy, they never allowed me to go out and purchase. Instead my husband had to go out to purchase things but the things purchased by the vendors did not satisfy me. Sometimes they purchased wrong items and made me angry.*"

SM33 shared, "*During surrogacy, I missed my children's proximity and family members. During surrogacy, my son was affected by chicken box but at that time, they never allowed me to see my son though my son who was desperate and longing for my presence and warmth and care at that time. But I could do nothing to him. In school, he got some severe injury and for that they didn't allow me to visit my home. Most of the time, I felt like I was being house arrested and failed to be on my own.*"

From the above three anecdotes it is clear that the freedom was totally lost and surrogate were under the control of the intended parents or the agents. The mobility restrictions, no permission to see the wards who suffered severely with illness, no freedom to purchase the food items or vegetables, etc, certainly indicated that surrogacy was also imprisonment for women for the 10 month period, whatever may be the kind of support money is could buy. Moreover, the surrogate mothers might have been convinced or they were forced to convince themselves, for two reasons that it would help them with decent money to attend to the emergency needs of the family and at the same time, it would help an infertile couple to have a child, which one cannot just equate with money. But in the process of commercial surrogacy, the women's body was seen only as an object and all restrictions, and stipulations were imposed on surrogates, only to get the baby healthy, least bothering about the surrogate mothers. Market dictates that everything was done for money and surrogate should accomplish that, without making anything an issue. It is not ethical nor humanistic because surrogates rendered such a noble service, and hence surrogates deserve decent treatment.

In everyday life, women do many things and they are always multi-taskers. But in surrogacy, women became inactive. They were never allowed to do routine work. The surrogates, who stayed out of home for surrogacy, mostly missed their children and few of them thought that they were house arrested. During surrogacy, surrogates were not allowed to do physical activities but they were asked to consume healthy and nutritious food. The inactiveness and the healthy nutritious food made women obese. During surrogacy, their physical and mental activities were restricted. Surrogates were forced to sleep in day time. These restrictions were imposed for the health of the child and nobody cared for the lack of freedom of movement and the pain caused to surrogates.

e. Feeling During Surrogacy Pregnancy

Each and every surrogate felt differently, during the surrogacy and also about relinquishing the child after delivery. Some surrogates felt good on some occasion but a few of them felt bad and suffered. There were different views expressed by surrogates, who shared different kinds of emotions and feelings.

SM08 shared, "*During my surrogacy pregnancy, I felt like being tied up with a heavy agreement. I thought that no liberty will be given to me because of the restriction imposed by the intended mother. The intended mother often forced me to follow the rules for the welfare of the child. But most of the time I could not oppose the rules, when I thought off disowning and I compromised myself.*"

SM05 and SM10 said, "*During pregnancy, I felt the difference in the experience of bearing the child of my own and bearing it for surrogacy. In the case of my own child, I never followed the standard rules and I did routine household duties. Compared with own pregnancy, in surrogacy pregnancy from each and every one cared me*" *(My husband, Association, Intended parents, hospital staff).*

In surrogacy pregnancy, each and every surrogate went through a different experience. It should be noted that those surrogates, under the care of intended mother or their relatives, received some kind of bad experience but surrogates, who were with their family, reported good experience. The reported feeling and experience of surrogates depended on the care and supervision of intended parents, association's representatives and doctors. For example, SM07 said, during surrogacy pregnancy "*I felt like deliver the child well and feel good*" and SM11 said, "*I always had touch to my stomach and feel a vacuum as the child is no longer with me which is again nature.*"

The mixed kind of experience and feeling revealed that surrogate mothers felt a sense of vacuum. It resulted from the feeling of not being with the child, after 10 months of hard labour, and it forced them to feel that the child was no more. There was not much difference between such mothers who lost their just born babies and the surrogates and it was an extreme form of emotion that the surrogate had expressed. This happened despite the informed choice that they must be detached from the child.

This experience of surrogates encouraged them to go for second surrogacy, even though some felt discriminated and this experience of abuse discouraged some from becoming a surrogate, second time. The good experience alone would have made the surrogate to accept for second baby and more the surrogate mothers opt for such offered only for meeting the economic needs. Though later they changed the mind that they offer to help a childless to have a child, the treatment meted out in the first surrogacy did not encourage the women to opt for a second surrogacy.

f. Comforts in Shelter Home

Surrogacy is a process, which is regularly monitored and reported and attended by the doctors. To undergo surrogacy, women migrated from their hometown. While practicing surrogacy, surrogates were asked to stay in a shelter home, rented house or their own house. In shelter homes, surrogates were under the supervision of doctors and hospital staff and in rented house, surrogate family also may stay or surrogates stayed with the intended parents or their relatives. During surrogacy, surrogates were in comfort zone, regarding food, care and treatment, love, sophisticated environment, immediate health care and special treatment etc.,

According to SM02, "*While practicing surrogacy in a shelter home, I received excellent personal care treatment, regular nutritious food, no household activities, health assistants' support, love and affection, immediate health care facilities and attention of the doctors, provision to continue to have hobbies, worship and watching television, support from doctors, commissioning parents, peers and special treatment which I may never experienced in life.*"

According to SM33, "*while practicing surrogacy, I was with intended parent's relatives. I did not get much comfort in a shelter home but I got special treatment from doctors which I never experienced in my life.*"

SM 12 said, "*While practicing surrogacy, in hospital, I received excellent personal care treatment, regular nutritious food, health assistants' concern and support, love and affection and complete relief from day to day quarrels from my husband.*"

While practicing surrogacy, surrogates were accommodated in three different places. These three different accommodations provided different kinds of comforts to surrogates. Compared to these three modes of stay, surrogates under the care of Associations, were in the comfort zone. Surrogates, who were accommodated with intended parents or their relatives, did not get many facilities. Surrogates with intended parents, had to do household activities like cooking, cleaning etc. It clearly revealed that intended parents or their relatives, abused and discriminated against the surrogates and did not follow the ART guidelines. Hence it cannot be altogether denied that surrogate mothers were ill treated and emotionally disturbed. The anecdotes related to such experiences revealed that except for a very few surrogates, the experience was quite good. This was due to the fact that at least for the sake of getting the child good, it was medically advised that the bearing mother must be happy during the pregnancy. If the surrogate mothers were treated well with all comforts, by all the stakeholders, the issues against the rights of the surrogate mothers may not arise. However, the sample size was too small to generalize the experiences of the surrogates in the universe.

Issues and Challenges Faced by the Surrogate Mothers

The surrogate mothers gave their consent after hard negotiations, self convincing, convincing husband, children, parents and others. Even after such convincing, it was not a smooth sailing for many surrogate mothers. They faced issues with family, community, agents and intermediaries etc. They are described below.

a. Issues in Family

While practicing surrogacy and after surrogacy, surrogates faced various issues in family and society. In society, the notion of practicing surrogacy was deemed immoral. But in Hindu and Christian religion, surrogacy is accepted.

The perception of society, regarding surrogacy, was seen as baby selling and equal to prostitution. Because of this perception, surrogates were not ready to disclose that they were practicing surrogacy or practiced surrogacy. In such a scenario, it is important to know what kind of issues a surrogate faced in her personal life, such as from her husband, in-laws and children.

According to SM05, "*In the context of misperception by of in-laws, equating surrogacy to sex work, myself and my husband planned that practicing surrogacy should not be disclosed to in-laws and relatives and avoided contacting them but a few months later, they got informed from someone. At that time, my in-laws abused me verbally. They mentioned that I was doing some illegal work (prostitution) and that's why I did not meet and have contact with relatives.*"

SM29 said, "I *got entire support from my family members but during my surrogacy pregnancy, I missed my children and children missed me for a long period. Certain behavioural change had happened. After surrogacy, when I back home my children did not come to me. It took much time for me to get my children back to me.*"

According to SM08, "*In my family, during and after surrogacy, I faced various issues like misconception by in laws, wrong notion about surrogacy in the society, children missing me for a long period and certain behavioural changes. In post surrogacy, I was verbally abused by the family members, health complications and no compensation for post surrogacy health hazards.*"

Majority of surrogates did not face any kind of issues in their family and the society because secrecy shrouded surrogacy. Surrogates were practicing surrogacy, with their husbands' or their parents' knowledge. A very few surrogates did not get any co-operation from their husbands because of the patriarchal notion. Surrogates concealed information from other people, who were close to them because of the wrong notions about surrogacy. Majority of surrogates migrated from their home town, for practicing surrogacy and some of them migrated, after surrogacy, to avoid ill treatment from the society. The intended parents provided money only for relinquishing the child but their health and negative reaction from society, was not taken into consideration. Even if only a few surrogate faced post surrogacy issues, it cannot be ignored as it did affect their personal life and hurt their own emotions, particularly the relationship with children because surrogate mothers agreed to the proposal only for the sake of welfare of the children.

b. Issues faced in Shelter Homes by Surrogates

During surrogacy, surrogates stayed in shelter homes, where surrogates faced various issues from doctors, nurses, inmates, intended parents or their relatives and intermediates.

According to SM08, "*During surrogacy pregnancy, I faced issues from the intended parents such as verbal teasing and ridiculing, ill treatment, quarrelling, not able to be relaxed and normal, and felt being house arrested. Also, forced me to sleep the whole day and every day, there was controversy over every activity.*"

According to SM33, "*I faced verbal abuse from intended parent's relatives, who did not allow me to consume healthy foods, sometime forced me to eat more, denied the freedom to speak with family members, each and every minute, indirectly teased me and hurt me and the thing which I did not like forced me to do that.*"

In shelter homes, compared to other surrogates, who stayed with intended parents or their relatives, some faced various issues while remaining surrogates did not face any issues in the shelter home. It clearly indicated that intended parents or their relatives never considered the surrogate's sacrifice and her health. The intended parents only cared for their child and showered verbal abuse on surrogates. This was due to the fact that they carried baby only for money. The ART guidelines do not mention about the supervision by intended parents and this is necessary to safe guard surrogates.

c. Psychological and emotional issues faced by Surrogates

During surrogacy, surrogates faced different kinds of psychological and emotional issues because they were away from their homes and children, till the delivery.

According to SM08, "*While practicing surrogacy, I faced pain because I was away from home for ten months and exclusion; also because I experienced abuse from in laws and relatives it had further aggravated my emotional imbalance.*"

SM09 said, "*It was difficult to convince my family members about the scientific process of surrogacy.*" SM01 said, "*To some extent I faced burden due to bearing child for others.*"

Majority of surrogates did not face any serious mental issues because they stayed with their husbands and children. Only a very few of them faced some issues, to some extent. The surrogates, who stayed away from home, faced the pain of being detached from home and some of them felt excluded or also faced abuse from the in laws. Thanks to the comfort being given to surrogate mothers, by all stakeholders, they were saved from psychological illness.

Relationship with Family Members of the Surrogates

a. Relationship of Surrogate Mothers with Husbands

Entering into surrogacy, it was mandatory to get the permission of the husband and without the husband's or parents' consent or permission, nobody can enter into surrogacy. For women, it was not an easy task to get permission from their husbands. When women shared about surrogacy, it was immediately rejected by the husbands because they thought that it was something wrong and also the patriarchal tradition did not allow them to accept surrogacy. However, surrogates themselves talked to their husbands and explained the process and all other arrangements. Also, they took their husbands to the physician and mediators, to clear their doubts. In some situations, where the husband was reluctant, the agent or the doctor talked to them personally, informing them about process and the noble service, that the women would render for the childless couple.

SM11 said, "*Very first he started to beat me because of surrogacy discussion. Later, he was also thinking, how will it happen? He didn't know. So I told him that everything will happen as directed by the doctor. Also I explained to him that many women suffered because of this and I am going to help childless couple and fulfil their desire.*"

SM9 and SM10 said, "*At first, My husband did not agree. He found it's hard to believe. We had major conflict. Later I explained to him that it is [done] through injections and not through direct sexual contact. There will be no problem.*"

In many cases, the surrogate mothers responded that their relationship with their husbands, in the initial stage (after they took the decision of surrogacy), was smooth. The reason for the same was the joint decision of the surrogate mother and her husband, to undertake surrogacy. Surrogate relationship, with their husbands, remained harmonious during the final stage of pregnancy.

Husband's consent and signature were compulsory, to undergo the surrogacy process. The anecdotes explain and conform to and strengthen the patriarchal notion of pregnancy, to be strictly within the limits of marriage and therefore, the husbands should a party to the negotiations. The husband also must sign the contract and not merely to give their consent verbally. Doctors or hospitals want to ensure that no issues or objections can be raised by the husband or families later on. Thus, the women, whose husbands opposed the agreement, were unable to take up surrogacy, thereby imposing a limitation on the agency and decision-making ability of the surrogate. Patriarchy was always a stumbling block to women's freedom, which is very well brought out in this study.

b. Relationship of Surrogate Mother with Children

Surrogate and her relationship with her children, remained important till the end of surrogacy. Majority of surrogates concealed the information from their children because if they came to know about it, they might suffer a lot. A few of the surrogates, who had kids below five years, did not have this problem but at the same time, one cannot estimate how much the children must suffer, parting with their mothers though there were members of the family to support surrogacy.

SM3 said, "*Very first, he did not know about my pregnancy but few months later, he came to know about it. At that time, we explained to him and he was convinced but at the end, he said, here after you should not do it again.*"

SM4 said, "Initial *days, my daughter did not know my pregnancy but after a few months, she came to know about my pregnancy. Immediately she started crying because she did not like the kid. She thought the kid will share her care and love from her parents and she considered the surrogate child as a competitor. Later, we took her into the association from association, she got the clarification and this is not our child and all.*"

Children are quite possessive when it comes to mother, sharing her love with others. It is natural and cannot be questioned. But convincing them is quite a challenge and if they were not convinced, they could end up with psychological illness, due to parting with mother and also bearing a child, which is not going to be theirs. It may turn into a big issue, where the very purpose of accepting surrogacy for the sake of welfare of the children, would be lost. That is the reason why majority of surrogates concealed surrogacy from their children because if they came to know, that would disturb and affect them mentally.

SM29 mentioned that "*She informed her children that she was going to pay attention to some old age people, for some days, as a service.* SM11 said '*I have informed my children and got their permission.*' *She also shared that* '*every year, my children would remember the birthday of the surrogate child and celebrate.*'"

It is interesting to record emotions of surrogate mothers and their maturity to accept the surrogate child, by remembering to celebrate surrogacy baby's birthday. Her emotions towards surrogacy baby were kept within the healthy limits with family members and the children too. This healthy different experience, should appreciated.

Surrogate Payment Receipt and Utilization

a. Payment received by Surrogate Mother

Surrogate payment is the most important factor, to become a surrogate mother. Many surrogates faced emergent needs and financial problems, that their families were unable to address, with their usual sources of income or assets. However, it is very important to know whether the surrogate mother received the contracted amount or not.

SM1 and SM2 said, "*I didn't claim amount but they fixed 3 lakhs for single child but due to twins I have received one more lakhs as extra. I have received money as per the agreement.*"

SM29 said, "*I demanded Rs. Four lakshs but intermediates said for single surrogacy compensation, it is only Rs. 3 lakhs so can't give more than that. After the delivery, I received Rs. 2Laksh and one month later, I received another, Rs: One lakhs (received in two instalments 50,000).*"

SM11 and SM12 said, "*as per the agreement, we should receive Rs. 2Lakh but we received only Rs. 1.5 lakhs and the remaining amount was cheated by the broker. We went to collect the balance amount from the clinic/doctor but the doctor said the balance amount was collected by the broker, so we went to collect the amount from the broker. But the broker woman shifted her home somewhere and we were not able to trace her and get the balance.*"

SM33 said, "*In my case, the doctor fixed the amount. The fixed amount was Rs 2.5Lakh for the surrogacy. But after the delivery, they did not give the amount. Before pregnancy, I had asked the amount for some urgent financial needs at that time and they had given Rs. 35,000 and then every month, they had sent Rs.3, 000 to my family. Like that, till the delivery I had received Rs. One lakh and the balance amount was not given and after three months, a few days before I called them to collect money, they said now we don't have money but will give later.*"

It may be mentioned here that payment to the surrogate mothers, raised some doubts. The amount should be fixed by the surrogate mother because surrogate alone has the right to fix the amount. The Researcher wanted to examine the gap between the amount quoted by the clinic or the hospital authorities, and paid by the intended parents and the amount surrogate mothers received at the end. The payments, to the surrogate mothers, were randomly decided by the physician of the clinic or hospital in all cases. It was found that these clinics or physicians blamed each other, for the rise in the amount, paid to the surrogate mothers. Similarly, there was no clarity in the contract about the payments, in case surrogate mothers were pregnant with twins. It is a normal practice that when the doctor found out about twin pregnancy of the surrogate mother, they consulted the intended parents, who, in most cases, were happy to be parents of twins and wanted to continue with the pregnancy. The deal was then forwarded to the surrogate mother, during four to five months of pregnancy, when she was not in a favourable situation, to bargain with the physician, for extra payment for the second child. Logically speaking, payment should be doubled for twin children and in this case, she was paid just one lakh more. Hence it was not a major gain for the surrogate mother. Majority of surrogate mothers received the compensation, at different points of time (through instalments), during and after the pregnancy.

Surrogate mothers thought that their lives would change and improve their economic status in the society. But the hypothesis went wrong. Majority of women became surrogates due to huge debt and after the surrogacy, surrogates spent maximum amount to repay the loan or pay for children's higher education or money for their daughter's marriage or other economic reasons. The driving factor, for a woman to become surrogate, was economic and the Researcher could not find any significant change in the economic status of surrogate mother, in the study area. Of course, they were getting little monetary benefit and their children could have two square meals for some days and they stayed in a house with asbestos roof to their huts but that is all. Also, few months later, again the surrogates were facing financial difficulties and they were again caught in the debt trap. Therefore, it is time to rethink whether the surrogacy arrangements really benefitted the surrogate mothers.

Another issue, that needs to be highlighted, is whether the payment received by the surrogate mother, actually improved the socio economic status of her family. It was found during the study that during the surrogacy process, while the surrogate mothers were in shelter home, the payment made to them, either in instalments or the entire amount was siphoned by their husbands, to spend it on alcohol or use it for setting up business, which, in most cases, did not flourish. Thus, when the surrogate mother got back to her house from the shelter home, after the completion of the surrogacy, she had little money left to take care of her children and herself.

The payment was quite decently done for majority of the surrogate mothers, but in certain cases, it was not paid as agreed. The fact remains that surrogacy market does promote exploitation by the intermediaries. Further, the amount fixes was not uniform. It ranged from Rs. 4 lakhs to Rs. 2 lakhs and the future payment ranged from Rs. 50,000 to Rs.1 lakh. There was also communication gap where the doctors made the payment to the agents, which should was done in the presence of the surrogates. Some woman broker was reported to the surrogates of their full payment. There has to be some inbuilt arrangements to make the payment in the presence of all the stakeholders to maintain transparency to prevent the cheating. The intended parents must be informed about the amount paid to the surrogate mothers and at least over phone, the intended parents must confirm the payment to the surrogate. Moreover, there is a responsibility on the part of the commissioning parents that they should do justice to the surrogate, who rendered her hard labour and noble service to them and if the payment was not made as agreed, might create disturbance to surrogates, which should not happen. It also further indicated that the surrogate mothers' ignorance was being exploited and they were rendered helpless and no one came to the rescue. Surrogacy leads to multiple exploitations and it also leads to subordination of women and, suppression of women.

b. Item on which Surrogate Payment Utilized

For each and every surrogate mother, payment for the surrogacy was the top most priority as they had accepted to the proposal for meeting the necessities. In the initial stage, intended parents did not make the payment to the surrogates. The intended parents spent only for the medical procedure undergone. After the confirmation of pregnancy, the agreement was signed. In majority of cases, the money was paid after the delivery only. In some cases, surrogates needed financial provision for their emergency and a portion of the amount was paid by the intended parents and the remaining amount was paid after the confirmation of pregnancy. After relinquishing, the surrogate mothers received the full payment, from intended parents, through agents or doctors. The amount could be paid in one instalment or a maximum of four instalments. The amount was used for various pressing and essential needs such as repayment of loan, children's education, purchase of house, medical expenses, etc.

According to SM16, "*I have received Rs: 1.5 lakhs compensation for surrogacy pregnancy after relinquishment. We had a small house but it was sold by my husband because my husband was addicted to alcohol. He consumes alcohol every day and for that, he needed money often but he did not have any property other than the small house. To fulfil his desire, he sold the house. I used the compensation amount for repurchasing our house, which was sold by my husband.*"

It Shelter is a basic need and alcohol consumption had made the household to lose even the small shelter to live. It is quite heartening that the surrogate payment had helped in redeeming the house though it was sold and thanks to the hard labour and emotional imbalance etc, which the women experienced, but at the end of the day, it could be rescued.

SM09 said, "*Doctor said my son was having severe problem in heart and for his recovery we should do major operation but we did not have such amount for his operation. Hence, I decided to undergo surrogacy. After surrogacy pregnancy, I used the entire surrogacy money for my son's heart operation.*"

It is equally appreciated that the surrogacy payment was used to save the life of son as it helped in the case of, in repurchasing a house. In fact, it needs to be appreciated that the surrogate payment had come to rescue the sample households. In other research studies (SAMA, 2006) majority of surrogates used surrogacy pregnancy compensation to repay a loan. A few of them used the amount for family purpose like children's higher education, savings for future, to start a business for her husband or to send husband for a job in abroad etc. The interesting thing in this research was that all surrogates before surrogacy pregnancy, had donated their eggs for their immediate financial needs. Hence the Researcher wanted to examine the surrogacy arrangement and the physical pain the surrogate mothers had to endure, during the process and whether the money was worth the trouble. A surrogate underwent surrogacy pregnancy for her family's and not for her pleasure. It was proved that for her family well being, women were ready to sacrifice their health and the patriarchal reproductive technology reinforced women only as child bearers. Surprisingly enough, all the payment was well utilized for productive purpose, which would have multiplier effect in future and safeguard and help in their future requirements.

Experience after Delivery

a. Referred to friends and relatives to practice surrogacy

Surrogate mothers were asked whether surrogacy was referred to any friends and relatives, encouraging them to practice surrogacy. Majority of surrogates reported that they did not refer to friends or their relatives because surrogates did not disclose their experience to relatives and friends. If referred to them, they would ask too many embarrassing questions. They would also doubt the absence of participants during surrogacy.

SM02 said, "*While I am practicing surrogacy, except my husband, nobody in my family, friends and relatives knew about this. I and my husband were not interested in sharing about my surrogacy contract with anyone. If we conveyed to others, they may misunderstand and treat us differently. Also, my husband may be teased by some other women or men in my native place. It would mar my husband's reputation also.*"

Many surrogates did not reveal the truth to anyone in their family, relatives or neighbours, except to their husbands. They even concealed this from their children also. Only one or two surrogates informed their children while practicing surrogacy. The hospital authorities or the association members of the needed the permission from the surrogate husband only. Those, who were separated or widows, needed the permission from their parents. In this scenario, all the surrogates concealed the information, regarding surrogacy. They were forced to hide the surrogacy contract even to family members due the stigma attached and the misconception surrounding surrogacy.

b. After Pregnancy Experience- Relinquishing the Child

Relinquishing the child was the most important part in the entire surrogacy agreement. In some cases, the surrogate mother became so emotionally attached to the child that parting from the child was painful. In majority of the cases, the decision, regarding the handing over of the surrogate baby, was made by the doctors and in a few cases, by the intended parents. Based on the growth of the child, the doctors advised caesarean operation in all the cases but in a few cases, the intended parents followed the advice of astrologers.

SM29 said, "After *25 weeks of pregnancy, based on the report of the ultrasound, the junior doctors sent report to the senior doctors for consultation. After that, senior doctor decided the time and date of the delivery. After caesarean, the doctor immediately handed over the child to the intended parents.*"

Relinquishing the child was decided solely by the doctors. Doctors and intended parents never waited for any permission from the surrogates and they were generally indifferent to the feelings of surrogates. When the doctor handed over the child, the surrogate was unconscious. If the surrogate were to be conscious, the surrogate may be emotionally disturbed during handing over the child. This might be the reason for handing over the child immediately. The baby was denied and the breast milk from day one onwards. The fear that the moment the surrogate mothers saw the child, they may find it difficult to part with or may suffer psychologically, was the major concern of the stakeholders. The commissioning parents were also concerned about their child and they took care of it right from the minute one onwards, and established the ownership. The surrogate just can not claim any ownership nor express the desire to see the child.

Benefits of Surrogacy

The Researcher wanted to know whether apart from the payment towards surrogacy compensation, any other benefits were received from the intended parents because relinquishing the child was not an easy task and money was not good enough reason.

SM05 said, "*After delivery I have received clothes from the agent and intended parents sent clothes to my family through association.*" The surrogates received some other gain from intended parents but important thing was that the intended parents were foreigners. Compared to Indian intended parents, foreigners provided some goods for the well being of the surrogate family.

The study revealed that many surrogates did not receive any other benefits other than surrogate payment. Though payment alone was the prime reason for accepting surrogacy, there could relationship post surrogacy but no intended parents wanted the relationship to continue. Many surrogates thought it was an incredible opportunity, to help someone, to create their family and find happiness. The surrogates did not expect anything more but just the happiness of the commissioning parents.

Awareness of Medicalization

a. Medical Procedures

It is important that while a few surrogates were not informed about the caesarean operation and the transfer of embryos, relatively more number of surrogates were aware of the fact that their own ovum was not used. Thus, while selective information was given, according to the convenience of the clinics and the parents, there was general lack of information about the authentic medical process, which the surrogates had to undertake.

b. Awareness of Tablet and Injection

During the entire surrogacy process, to confirm pregnancy, surrogates were advised to take plenty of tablets and injections. To know whether the surrogates were aware of the tablets and injections, they were asked whether they were informed about the tablets and injections. It was understood that not all the surrogates were aware of the tablets and injections given to them. They reported that doctors did not disclose anything about the use of tablets and injections. Compared to natural pregnancy, under technology supported pregnancy, to get positive results number of tablets and injections were prescribed by doctors, for the well being of the child for the intended parents but doctors did not consider as the health of the surrogates. The contract between the two parties was considered as sale contract and this exchange was deemed to be commodity trade. The commoditization prompts the trader to ignore the health of the surrogates.

Role of Clinic in the Surrogacy Process

a. Medical history and Investigations or Examinations

Before initiating the surrogacy process, women were asked to undergo various medical screening, especially obstetric history was mainly considered and several tests had been conducted to ascertain their 'fitness' for carrying the pregnancy. The experience of childbirth (their own) was very important for the surrogate mother, for entering into the surrogacy process. The fitness of the surrogate's body for surrogacy, was evaluated, by a thorough medical examination and by various tests, to check the hormone levels and uterine lining, and to make sure that they did not suffer from any disease that would render them 'unfit' for the surrogate pregnancy.

SM33 said, "*I should already have children by normal delivery. And I should not have any medical problem. To ascertain my medical fitness they do master health check up like blood, urine test and uterus test. My uterus should be in good condition and if I have cyst or any other problem, the physician will reject me. Also, physician asked my family medical history.*"

SM11 said, "*I have already undergone surrogacy once but I tried to undergo surrogacy for the second time. For that I went to see the infertility physician but after the medical examination, the physician said there was cyst in my uterus and she asked me to undergo treatment. Now I am under treatment.*"

Thus pre-surrogacy involves a lot of medical examination and offering oneself to surrogacy was not an easy task, leave alone the bearing of the child for 10 months. All possible tests were done without considering the pain it brought to the surrogate but only concern was the birth of the surrogate child the well being of the child. A few surrogate mothers reported that the following tests and diagnostic technique were done: blood test, urine test, and ultrasounds. They explained these tests and techniques in different terms but never, with any surety, about the purpose or need for them. While surrogates were made to undergo several medical tests, they were rarely informed about the purpose. Many surrogates, interviewed for this study, were not able to explain either the number or the purpose of the tests conducted, beyond a minimum level. Mostly, all had shared that the "*Tests were done to check whether there is any disease and ultrasound was done to check whether there is sort of problem in the uterus.*"

SM5 said, "*In the beginning, blood and urine tests were conducted and everything was fine in the report. Later, they started further process but my blood group was negative. Initial days, they asked me to take some treatment separately till the delivery physician asked me to take a few tablets to avoid complications till delivery.*"

In a few cases, women were found to be suffering from deficiency like anaemia and the physician prescribed appropriate medicines. They were subjected to further medical interventions to confirm that their bodies were 'ready' for further processes and procedures. Thus before the surrogacy process, the physician administered complete master check up because they had to give cent percent positive result, to intended parents, who spent a lot of money. Interestingly, there was contradiction in surrogacy process even at a very primary level. The investment in making the surrogate's body 'healthy' was, therefore, accompanied by its own politics wherein all efforts were geared towards the ultimate aim of ensuring the birth of a healthy baby, without any complications, without caring too much for the welfare of the surrogates after the delivery. Utmost care was taken to keep the surrogate mothers healthy without any major complaint with the uterus, because otherwise it would be risky and the commissioning parents also spent a lot of their earnings for getting a biological child. Such concerns were justified as any failure to get the child would be a huge loss to the intended parents and physical pain for the surrogate mothers, which one cannot compensate.

b. Egg Donation

In this study, many surrogates had donated their eggs before surrogacy, once or twice. The intermediaries first approached the women for egg donation and later persuaded and convinced to accept them surrogacy. Women did come to know about the entire surrogacy process, after the egg donation process. Before the egg donation, women entertained doubts about surrogacy like whether practicing surrogacy was good or bad. The study revealed that, *surrogate egg was not used in surrogacy process and surrogate was used as a carrier. For surrogacy process, they used eggs from intended mother but in a few cases, the intended parents borrowed egg from some other women such as egg donors.*

The doctors also tried their best to give the intended parents egg but on repeated failure, the egg donors came to the rescue. But surrogate mother's egg would be tried first.

c. *In-Vitro* fertilization

The IVF procedure was also carried out in the traditional surrogacy but in commercial surrogacy, IVF played a vital role. For IVF, semen and eggs were generally harvested from the intended parents by injection and never harvested from the surrogate mothers, except in a few cases, when the intended parents demanded it. In this study, surrogate mothers donated their eggs elsewhere and not for surrogacy (Jadva et al, 2015). In surrogacy, the fertilized egg would be transferred to the surrogate, through a process called Embryo Transfer (ET). SM13 said, "*Embryo transfer was made by the doctors but I did not know whose egg and semen were used for the* IVF."

This is the medical secret. Only the Doctors would know whose egg and semen had been transferred. If the egg or semen, donated by the intended parents, happened to be weak, in order to have the conception, the doctors would resort to other strong egg and semen. But there is no guarantee that egg or semen alone has been from intended parents alone would be injected. The extreme concern of everyone involved in the process was to have a baby and kept at the Centre. Almost no information was provided to surrogates, about the procedures or about the techniques that they were subjected to. This information was shared by the physician only with intended parents. In fact, even if they had used surrogate egg, surrogates may not know such facts. This was evident from the fact that the flow of information was heavily influenced by the interests of intended parents and a successful birth and smooth relinquishment prevailed over any considerations of ethical protocols or of informed consent from surrogates. It only indicates that the surrogate mothers were kept always at the receiving end where the market was keen in keeping them away from knowing any information they were kept in the dark. She was there to just hire out her uterus and she had to co-operate fully till the delivery. Nothing else was necessary nor warranted. This was how the entire surrogacy process was arranged where agents, doctors, intended parents alone were informed.

d. Restricting Sexual Activity

During the surrogacy period, the surrogate mothers were absolutely made to restrict their sexual activity. Their sexuality was under complete control. To enter into the surrogacy procedure, the surrogates were advised by the physicians to completely stay away from sexual relationship with their husbands. It is one of the most important norms the surrogates should follow till the delivery. The rationale, provided by the doctors and agents, was that after the transfer of embryo, to ensure that it was successfully implanted, the surrogate should not conceive through her husband.

SM 05 said, "The *doctor and mediators informed me that once you signed the agreement, you should not involve yourself in any sexual activity with your husband*."

The restriction was imposed to avoid transmission of any disease. In order to ensure self-denial and to avoid problems, that surrogate may face in the event of conception and pregnancy, surrogates were asked to stay under the care of intended parents, mediators and doctors. The control over the surrogate's sexual behaviour was motivated, by the concern of both the intended parents as well as the doctors, to ensure conception only through the transfer. From SM5's account, it was evident that the overriding concern was to ensure successful conception and greater success rates. In other words, to deliver a healthy child, surrogate should strictly refrain from sexual activity. Though it is medically advised, it is against nature and the urge to sexual need is totally controlled, which also is included in the price fixed for surrogacy as an invisible cost. Surrogate mothers must ensure that their husbands do not develop into extra marital affairs, particularly for young couples refraining from sex for 10 months, would be really a challenge and it requires self control. If the husband were to seek commercial sex worker, the entire life of the surrogate would be ruined and the very purpose of surrogacy would be defeated. So it is all the more challenging in the patriarchal context that the surrogate mothers must take a decision to be away from family, husband, children and hence the surrogates were forced to be vigilant. Such forced seclusion may also invite psychological problems. In other words surrogacy is not just an easy task and given an opportunity, they may lose their own husbands because they stayed away from the family for a year or so.

e. Regulating Diet

For the healthy growth of the foetus and to reduce complication during gestation, the surrogates were asked to eat specific nutritious foods, prescribed by the physician. Surrogates reported that they were particularly asked to frequently consume fruits, juice, curd, almonds, vegetables, coconut water and also asked to have only home cooked food. Especially all the surrogate mothers were forced to have protein biscuit or protein powder in milk.

According SM29, "*The intended parents forced me to have food often but due to severe vomiting, I could not eat. I could not have even a single glass of water and if I consumed water, then started vomiting immediately. But, she (intended mother) never cared for my problem and she forced me till the delivery. Also, they asked me to avoid papaya, orange like citrus fruits, oil, spicy, salty foods.*"

SM33 said, "*The first five months I had prepared food according to my wish. Later relatives of intended parents accompanied me and they always forced me to consume food. They prepare food for me which I never liked nor the way in which is it being prepared. It won't be tasty and hence I tried to avoid such food. Since I do not like such food and more than that the way in which it has been prepared but it is forced to be taken and end up with continuous vomiting.*"

From the above responses from the surrogate mothers, it is clear that the intended parents and their relatives recorded much concern about the health of the baby and hence they forced surrogate mothers to consume all nutritious food but unfortunately the surrogate's body did not agree with. As per the guidance of the doctor, regulated diet would be brought by the intended parents, to the surrogates. Some intended parents paid a separate amount, to purchase fruits, nuts, etc. A few intended parents sent all those things, through the agents or the relatives of intended parents. Also, during surrogacy pregnancy, all the surrogates till delivery, were compelled to take protein biscuit or protein powder with milk, for the well being of the child. But during pregnancy, it is not always possible for pregnant mother to consume all kinds of food though it is needed for the health of the child. Normally, any mother would force the pregnant woman to take nutritious food for the sake of the health of the child. But in surrogacy pregnancy, in addition to baby's health, money was also involved and any failure, would cause dual or triple burden- the baby, money spent and the suitable surrogate to bear the child. Hence market has its eye on the health of the baby and surrogates are forced to take such food, whether they like or not.

f. Restriction of Physical Activity

The surrogates are generally asked to take complete rest and reduce their physical activity to the minimum level. The surrogates were advised to sleep, in the day. Many surrogates reported that due to pills and injection, they were always feeling sleepy the whole day.

SM5 said, "*I stayed with my husband during the pregnancy. My husband did all the household activities and I simply took rest the whole day. During the entire pregnancy, I was never allowed to do any activities and I was always lying on the bed. I used to get down from bed only to go to the rest room. My husband took care of my child also.*"

SM08 shared that "*They forced me to sleep in the day time but because of injection pain, I could not sleep. Also, they forced me to read books and hear traditional songs. For namesake, I would see serials. Often the intended mother would scold me, asked me not to watch the serials. They never allowed me to walk and restricted me from physical activities.*"

Surrogates were also asked to travel to the clinic for appointments, in cars/taxis. SM01 reported that, "*The doctor asked me to come by car. It was 'twins' and it might be risky if I do not go by. So after five months, they [commissioning parents] also asked us to travel by car and they said that they would pay for it.*"

A few of the surrogates registered that it was almost a house arrest for them, where their movements were restricted strictly. Restrictions of physical activities made the surrogate to feel like they were sluggish. In the day time, surrogates were forced to sleep. As per science, human brain can be active in day time but during sleep, brain does not work. To avoid unwanted thinking, surrogates were forced to sleep. Otherwise surrogate's negative thoughts might affect the child in future. Respondent's surrogates reported that they had to completely surrender their routine and function as per the directions of the doctors. They had to sleep in the day time, which was against nature. These restrictions caused psychological illness surrogate mothers had to adhere to the norms.

g. Stable Supervision and Surveillance

The surrogates delivered the baby under the supervision of agents and doctors. Mostly, agents were asked to supervise the surrogates, to make sure that they took the routine medication, injection and behaved the way prescribed by the doctors. Also, household routine should be avoided till delivery. The only thing surrogates should follow was the routine like injection, medication, consuming nutritious food and taking complete rest and the remaining activities were strictly restricted by the doctors and the agents, for a healthy child gestation.

SM8 said, "*I was always under the supervision of intended mother. She observed each and every activity and forced me to avoid watching the television serials and forced me to read traditional books and hear traditional songs.*"

SM02 said, "*We (surrogates) were accommodated in a single home. When they (agent) provided food, we were monitored from morning to night. Often they would call and enquire whether we had medication and other routine, prescribed by doctors.*"

During pregnancy, surrogates received healthy and nutritious food. But restrictions made them think they were treated as bonded labour. Also, surrogates, who had one or two delivery, did not much care about the pregnancy as they had prior experience of how to handle pregnancy, but the investor (intended parents) wanted to take care of their child. But at the same time, they could not involve themselves directly because that would reveal their identity. Hence intended parents handed over the responsibility to the agents and doctors, bossed over the surrogates, though some of them were quite flexible and liberal with the surrogates. However, the surrogates who were under the supervision and surveillance of agents and doctors, reported that they were in comfort zones, compared to the surrogates, who were with intended parents and relatives of intended parents, who were ill treated, physically abused and threatened. When the intended parental control was not there, surrogate mothers enjoyed some freedom. The restrictions were designed to ensure safe delivery and get a healthy baby, but unfortunately, it deprived the freedom of the surrogates and they had to accept the loss of freedom for the sake of agreement, to receive the money for hiring out their uterus. It might be quite challenging but equally threatening to give up in between, though at times the surrogate mothers really did not feel comfortable with the treatment.

h. Embryo Transfer-Multiple Cycles

Normally, successful rate of IVF is not high and a single try does not give positive result. In surrogacy, the embryo transfer should be made by the doctors but the embryo transfer did not give positive result in the first attempt. For getting positive results, the doctors may make many attempts to get positive result. Except in the case of a few, many surrogates had to go through two or more cycles.

SM5 said, "*After the successful and complete medical test, a few days later, they started to do embryo transfer. But first time, it was a failure to me and then a few days later, doctors tried for the second time. It was a success to me and they asked me to stay for 15 days in the hospital.*"

SM1 said, "*After successful master check up, they started to do embryo transfer. But first time it was a failure to me. Only fifteen days later, I came to know that the embryo transfer was failure. We went back home. A month later, DNC was done and they started the second embryo transfer. Physician informed that due to depression first time it failed. Hence, they asked me to go home and then come back fifteen days later. Fifteen days I took complete bed rest and had injection and tablet for fifteen days. Later I went back to hospital and got the embryo transfer done which turned positive.*"

Many surrogates were not aware of the details of the process of surrogacy. They had only limited knowledge and they did not know about the details of the entire process. The physician never disclosed the details to the surrogates and all decisions, regarding surrogacy process, were taken by the physician and the intended parents. As the intended parents were spending a lot of money, to develop their genetic link, it was their huge concern. Thus there was pressure on the physician to give positive result. Because of lack of information about the possible effects of treatment and the IVF procedure on their bodies, multiple cycles were done to the surrogates, as a preferred convention, to make it a success but without mentioning the possible increase in the associated risks. As per the draft ART Regulation Bill and Rules, the surrogate should not undergo more than three IVF cycles. But it is known only to the Doctors as to how many times the surrogate had to undergo the IVF as the surrogates were completely ignorant and no informed choice was made while giving the medicines as to what purpose a particular tablet or injection was given. Surrogate just accepted as if it was all for the successful pregnancy and delivery as the surrogate also wished to be relieved from the bond at the earliest and go back home. In fact, failure of one embryo means more physical and mental pain for the surrogate. No extra payment was made. Hence every surrogate mothers was desperate to get the conception successful at one go or two. But the fact is that even if it failed for two times, the third try by the doctors may not be informed to the surrogate.

i. Multiple Embryo Transfer Cycles and Foetal Reduction

In the case of commercial surrogacy, the physician preferred multiple embryo transfer, to get cent percent positive result. Multiple embryo transfer may initiate multiple pregnancies but these multiple embryo transfers, may lead to multiple foetal reduction. Fifteen days after the transfer of the embryo, the physician checked the success of the embryo transfer, to find out whether the embryo was successfully implanted or not. Once the embryo was successfully settled, again the physician checked the embryo's growth. If the embryo's growth was on course, the intended parents could decide how many children they wanted, one or two. After the decision of the intended parents, the remaining foetus could be reduced by the physician. The possible risks of multiple pregnancies would be increased chances of miscarriage, obstetric complications, premature deliveries, and birth complications. Foetal reduction may cause bleeding, perforation, infection, premature labour, and loss of all foetuses. These, however, did not find place in any discussion, before surrogates underwent these procedures (Sama, 2006). Meanwhile, it was also found that in some ceses, as per the desire of the intended parents, the physician allowed two embryos to get twins, without the knowledge of the surrogate mother. In such a scenario, the surrogate mother received some extra monetary benefit, not much though.

SM1 said, "*The first embryo transfer cycle was a failure and a month later, the Doctor conducted a Dilation and Curettage (D&C). A second try was made for the second cycle. In the second cycle, I was not informed as to how many embryos they had inserted. Fifteen days later, went for scanning and after scanning, they said that three embryos' growth was very good. I felt so happy because of the first failure. Immediately I had doubt about how could I carry three children? But physician asked me not to worry because two was enough one embryo will be removed through injection. Then, they removed one embryo and after removing that embryo, I had bleeding. I thought bleeding might affect remaining embryo but good that it had not happen.*" *It is evident that SM1 was never informed about the transfer of multiple embryos nor about the reduction; her consent was never sought. D&C may cause cramping, heavy or prolonged bleeding, fever, pain, abdominal tenderness (web MD).Also the bleeding while removing one embryo was quite risky for her life but somehow sustained.*

*M*eanwhile, even the physician or the agents did not inform the surrogate, about the transfer of embryos nor about the feotal reduction. They never asked her permission. If they had considered the surrogate's health, they would have asked her consent but all decisions were taken by intended parents and the physician. Both had ignored the surrogate and treated her like a slave. It clearly revealed that once they entered into contract the freedom was lost and surrogates had no rights to enquire about their health. Till the delivery, surrogate health was considered, only from the point of the intended parents particularly health of the child. Even if the health of the surrogate mothers was taken into accounts it was only for the sake of the child.

SM33 said, "*The first attempt gave positive result and they said they had inserted two embryos but only one embryo successfully settled in me. Meanwhile, the intended parents blamed me that because of me, another embryo was spoiled.*"

Transfer of multiple embryos, done to satisfy the desire of the intended parents, induced various health hazards to surrogates. But when the uterus came under trade, the demand side never considered the hazards of the supply side. Even the basic human courtesy was missing on the demand side. They considered only the interests of the intended parents.

The Draft ART Bill – 2010 has mentioned that if the first embryo transfer failed in surrogate mother, she may, if she wishes, decide to accept, on mutually agreed financial terms, at the most two more successful embryo transfers, for the same couple that had engaged her services in the first time be permitted. No surrogate shall undergo embryo transfer, more than three times, for the same couple. But, when the physician or the intended parents obtained the consent of the surrogate mother, to undergo second attempt, what does the Bill say on that? If they undergo more cycles, it should be recorded but no such data were maintained. According to the provisions of the law, if the embryo transfer was done more than the twice it must be recorded. Hence it is not registered in general in any clinic. But it does not mean that the embryo transfer was not done more than twice, as in several clinics, no data or report has been maintained as the researcher observed. Also, it should be informed to surrogates, about the risks involved in multiple embryo transfer and gestations. But the surrogates generally did not get such information nor did they ask for it.

j. Multiple Injections and Tablets

After the IVF, to get positive results in surrogacy pregnancy, surrogates were asked to take a number of injections and tablets. During a normal pregnancy, a limited number of need based injections and tablets would be prescribed by the physician. But in surrogacy pregnancy, for the pleasure of intended parents and to increase the success rate of surrogacy pregnancy, surrogates were forced to take multiple injections and tablets, to get positive results. But many surrogates were not aware of the effects of the tablets and injections. They simply took the injections and tablets, as prescribed by the doctor. Except a few, majority of surrogates did not get any information, regarding the purpose of tablets and injections, as prescribed by the doctor. In this study, many surrogates shared their experience, on injections and tablets as follows:

SM33 said, "*After IVF, doctor advised me to take injection every day. For a full month (30days), the doctor put the injections on my hip and thighs. Also, as per the prescription, till delivery I had to take tablets every day thrice. I asked the doctor, the purpose of the tablets and injections and the doctor replied that, injection is for settlement of karyon in uterus and did not disclose the purpose of the tablets. Meanwhile, most of the time I could not sleep because injections are gave severe pain on my body. I could not tolerate the pain which I have experienced. I had delivered the baby six months ago but yet I had pain on my hip and legs. Also, I could not walk few minutes continuously and during walking, I suffered migraine. And because of the tablets every day, often I vomited and because of vomit I could not consume food properly due I have lost weight.*"

All the surrogate mothers reported body changes, after taking the injections and tablets. From the surrogate experience, it is evident that to get the positive results, for the benefit of the intended parents, the IVF surrogates were indirectly forced to have injections and tablets. The doctor did not disclose any information regarding the tablets and injection. Information, regarding whether the injections and tablets would have any side effects in future, was not shared or conveyed to the surrogates. In other words, surrogate health was taken for granted by doctors and intended parents because of the commercial nature of surrogacy. The Center for Bioethics and Culture Network (2014) reported that drugs, commonly used for women in commercial surrogacy pregnancies, did have side effects. The health risks, that the surrogates might face (as a result of experiencing IVF, injections and tablets), were not considered and her 'worth' was entirely dependent on a measurable reproductive 'outcome,' i.e. the baby.

The health exploitation of lower-income women, by higher-income couples, was clear in commercial surrogacy. In commercial surrogacy, women were not considered as a subject. Instead women and their parts were considered as objects and mother as the baby making machine. In this scenario, the feminists argued that to seek to protect women's reproductive labour, from the sphere of the market, was not adequate. The vulnerability and power relation creates the exploitation. There are no standard 'treatment' protocols for ART procedures and this paved the way for exploitation of surrogates physically.

k. Selection of Embryos and Sex

The practice of selecting embryos was part of the procedure, to gestate a healthy child. But surrogates reported that "*Embryo selection was carried out by the physician but did not know whether the physician was doing sex selection or not*" and if there was sex selection, it was carried out without the knowledge of surrogate mother. Mostly intended parents liked to gestate a male child. The sex selection was illegal but couples, who had either lost or did not have children, were especially keen on having a son. There is a belief that in their old age, they needed a son to take care of them and to rule their properties.

l. Caesarean - Controlling Birth

In surrogacy, all the gestations were, by force, caesarean because physician and intended parents were not ready to take the risk. Even a single delivery did not normally happen in surrogacy. Normal delivery could not ensure safe delivery for the child and the caesarean time was fixed by astrologers. Many reasons were behind caesarean such as multiple pregnancies, delivery complication, etc.

SM29 said, "*Before initiating the surrogacy process, they convinced me that it would be a normal delivery because I was afraid of caesarean. In the initial stage, I worked for normal delivery. But later I came to know that they never allow normal delivery and it should be always caesarean. When, they realized that I was scared of caesarean, mediators and doctors assured me of normal delivery. Then, they put injection on my hip side and they closed my eyes with a white cloth. After that, they carried out the caesarean. Meanwhile, I was half conscious during the caesarean but I realized the entire thing such as tearing the skin, taking baby out, stitching and baby weeping sound.*"

SM33 said, "*At first they informed me that delivery will be by caesarean only. But in the end, they said 90 percent chances are there for a normal delivery. Also, to do normal delivery, they put saline (trips) and injection. They waited the whole day, from morning 06.00am to evening 06.00pm for normal delivery despite my severe pain but normal delivery did not happen and finally it was caesarean.*"

In all the cases, many surrogates were frightened about caesarean because of severe pain. The irony was that the surrogate mothers reported normal delivery for their own children. The surrogates described the effects of undergoing a caesarean operation. SM1 and SM5 both had one or two children, born through normal deliveries and in their surrogacy arrangements, birth of the baby was through caesarean. After the birth, both experienced pain in the lower abdomen, near the sutures, for three or four months. Also, SM5 explained that' *if I had known about caesarean pain before surrogacy, definitely I would never have undergone surrogacy*. The forced caesarean caused multiple side effects, to the mother and the child, like infection, haemorrhage, thromboembolism, injuries in internal organs (NHS, 2015). These issues amount to violations of surrogate rights.

m. Breast Feed

Many surrogates were not allowed to breast feed the child because mother and child bond might happen because of the breast feeding. Under these circumstances, majority of surrogates were given injections to stop them from lactation, after giving birth. The surrogates were informed that they would not be breastfeeding the child. Out of the thirty three sample surrogates, who had given birth, no one was allowed to breastfeed the child and they were given medication in the form of pills or injections, to stop them from lactation. Some of them carried out natural remedies to stop lactation. Mostly, the intended parents did not permit the surrogates to breastfeed kid.

SM33 said, "T*hey are asked me to feed the baby but I did not have milk secretion and a few times they sucked milk with the use of pump.*"

SM29 said, "After the gestation, they asked me to feed the child but I could not feed the kid because I was afraid that emotional bond might be created because of lactation."

The decision, not to allow the surrogates to breastfeed, was aimed at preventing surrogates from developing any kind of emotional bond with the newborn and also to avoid the possibility of surrogate, claiming any rights over the child. After gestation, the child was immediately handed over to intend parents by the doctor and the mediator. This handing over was made easier due to lack of consciousness after delivery because of caesarean by injection. Besides, injections and pills, used to prevent lactation, had serious side effects, ranging from dizziness and nausea to hair loss, etc. There was also the possibility of breast enlargement, an extremely painful condition, which may arise if the woman, who was lactating, was not allowed to breast feed the child.

n. Post delivery care

After the gestation of the child, if surrogate faced any health complication, the doctor should give treatment to the surrogate. It is the responsibility of the doctor and the intended parents. But the surrogates did not know how to contact them. Compared to normal delivery, in the surrogacy process, surrogate took too much of pills and injection, to deliver a healthy child. Also, they were forced to undergo caesarean section. These circumstances could create some hazards to their health because as mentioned above, the pills and the injection may give side effects.

SM11 said, "*After delivery, few months later, I suffered some health problems like severe stomach and hip ache. I went to the hospital for treatment. I reached the hospital morning, but till evening, they never cared about me. I simply sat there till evening like am orphan. Nobody offered to help me. During pregnancy, they took care of me but ignored me post delivery. After surrogacy, I was affected by cyst in my uterus but I did not know the reason behind the uterus cyst.*"

Before going into the surrogacy procedure, doctors and mediators assured the surrogates that if they experienced any problem, after delivery, they may approach the hospital. However, in the case of surrogates, who had delivered, no steps were taken by the doctors or clinics, in following up the surrogates. It was usually the surrogates, who approached the doctors, in case of any problems, with or without the support from the commissioning parents, which varied in each case. In general, there was no arrangement for post delivery treatment. In case of any serious issue, if the commissioning parents should any concern, the surrogates might get some medical support. But the post surrogacy concerns were not the same as during surrogacy, which was well understood and hence they were prepared to face it.

Information about Intended Parents

a. Familiarity with Commissioning Couples

Knowing the commissioning parents, face to face, means that it would be easy for surrogates to know their intended parents' names and their language. Also, language was the barrier to surrogates, to share their feelings and their experience, with commissioning parents. Also, in majority of cases, mediators and physicians did not allow the participants, to meet their intended parents because they must have thought that knowing commissioning parents would lead to problems in future, for the intended parents and their child. During the interview, the question was posed to surrogates, about familiarity with intended parents.

SM5 *said* "*Mostly Associations, mediators and physicians never disclosed the details of intending parents with the surrogates and also, they maintained confidentiality.*" Meanwhile, SM29 stated that "*Some of the intended parents were eager to stay with us, to take care of their child and feel the growth of the child.*" Similarly, SM33 said, "*They will arrange their relatives to take care of the child.*"

The research details proved that all the sample participants did not know their commissioning parents before surrogacy. The intended mothers, especially foreigners, were eager to meet the participants before, during and after the surrogacy. But in India, majority of Indian couples avoided meeting the surrogates but the commissioning couples knew about their surrogates. At this point, the interesting fact was that if the commissioning parents and surrogates maintained relationship with each other, they would share all the things involved in surrogacy, including the cost for surrogate. Under these circumstances, the mediators could not cheat the surrogates and no wonder in India, that intimacy between surrogates and commissioning couple was not encouraged.

Some related Interesting Experience of Surrogates

a. Feeling Good - Pregnancy Experience

Surrogates reported two different kinds of experience in pregnancy such as her own and surrogate pregnancy experience. According to ART Bill, each and every surrogate should have a child, with normal delivery. To find out the best experience of surrogacy, the surrogates were asked and the response was quite surprising.

According to SM07, "*Compared to own pregnancy, surrogacy pregnancy experience was very good because in surrogacy pregnancy, I received care and love from every one such as my husband, intended parents, agent and clinic staff. During surrogacy pregnancy, I stayed with my husband and child. During pregnancy, my husband took care of me a lot but in my own pregnancy, I did not get this much of care from my husband. Also, during surrogacy, my husband did all the household activities like cooking, washing, caring of child etc. Also I consumed lot of healthy food during surrogacy pregnancy.*"

Regarding this question, the response was unexpected because the Researcher never thought of this perspective. At the same time, the Researcher wanted to examine the importance of this response. It is worth registering that surrogacy pregnancy experience was good for those surrogates who stayed with their husbands and shelter homes. But those, who stayed with intended parents or their relatives, did not give this answer because as mentioned above, the surrogates who stayed with intended parents or their relatives, were verbally abused. The husband of surrogate did not take care of her wife, in the birth of his own child but in surrogacy, due to the offer of money, he took much effort to bring a healthy child, to the intended parents. It has to be critically analyzed to understand how money played an important role and how it changed the power of patriarchy in the technological world. Commoditization broke the rules and power of the patriarchal society. Money did play a crucial role in human life and broke the cultural ethics.

b. Interest to Undergo Surrogacy Second Time

According to ART Bill, women from the 20-35 age group alone, could practice surrogacy. The surrogates, who practiced surrogacy for the first time, the question was asked whether they were willing to undergo the second surrogacy. Majority of surrogates were interested in surrogacy for the second time but not immediately after one year. But some surrogates were interested in second time surrogacy but due to age constraint, the doctors did not accept them for surrogacy. They were asked purpose for their willingness to practice the second time surrogacy. Majority of surrogates reported that they wanted to help the childless couple and bring happiness to their family. The surrogates, who may have economic vulnerability in their life and hence aspired for practicing surrogacy for the second time, differed totally from the first time surrogates. When they shared their first experience, their first desire was to earn money. Hence the Researcher wanted to critically analyze the desire for the second time surrogates because in the first experience, they got enough money. From that kind of money, the first time surrogate should be economically comfortable. But the study revealed that those who had practiced surrogacy the first time, received a decent sum of amount. But it was used for repaying their loan. However, many of them were again forced to get loan for the basic needs of the family members. Thus there was difference between the first and second time practice. The first time surrogate suffered very critical economic situation but in the second time, they found some way to earn money, to repay the debt and develop the family.

c. Aware of Delivered Baby

In commercial surrogacy, women were hired to carry and deliver a healthy child for others, on a commercial basis. After the success of *in vitro* fertilization, women were asked to take complete rest and consume healthy foods. Also, they were advised to take plenty of tablets and injections, for the well being of the child. After a certain period, the child was delivered by caesarean, without any concern for the surrogate. Each and every step of pregnancy decision was taken by the physician, without the knowledge and the consent of the surrogate. But after delivery, physician, agents and intended parents did not even disclose the sex of the delivered baby, to the surrogate. SM09 shared "*None of them disclosed the sex of the baby delivered by me and they took the baby away from me during my unconscious time and three months later only I came to knew that I had delivered sex of the baby, as a boy.*" Majority of surrogates shared that they were not aware of the sex of the baby they had delivered. But surrogates wished to know not only the sex of the child but also the growth and development of the child.

d. Desire to see the Infant

As per the agreement, surrogates were not allowed to see the face of the child after delivery. Meanwhile, all the deliveries were made by caesarean operation and surrogates were mostly unconscious after the delivery. But to know the feeling and wish of the surrogate, the question was posed to the surrogate. Majority of surrogates wanted to see the face of the child and they would like to know the growth and development of the child in future. Few of the surrogates wished the intended parents happiness, even after relinquishing the child.

SM05 said, "*I don't want to see the child face because if I see I cannot leave the child and it is better to me for not seeing the kid. Meanwhile, intended parents may share their feeling with us and if they can't meet us in person, means they can share through association representative or by call. If they convey their feelings, I will feel good and it will make me satisfied and its gives me pleasure.*"

SM33 said "*During pregnancy, I didn't think that it's some other's child and I have also spoken to the child during pregnancy. But after relinquishing, I would like to see the kid's face but they did not show the child to me but after relinquishing the child, the intended father prostrated and thanked me.*"

As per the surrogacy agreement, surrogates were not allowed to see the surrogacy child after delivery and in future also. SM01 said, "*Before entering into surrogacy, I strongly kept on mind like I am just going to carry the child only and also it's not my child.*" Doctor and the intended parents also did not wish to show the child to surrogates. It may be due to the fact that if the surrogate were to see the child, she may not leave the child and she may create problems. Though she was not genetically related to the child, since she bore the child, she might want to see the child and later may express desire to keep the child with her and may develop emotional bond with the child. Hence surrogates generally are not allowed to see the child and these things were clearly mentioned in the agreement, to safeguard the intended parents and the child from the surrogates. But many surrogates wanted to see the child after the delivery, because as a mother, who bore the child for 10 months, it was inevitable. However, it is observed in this study that in some cases, the intended parents did permit surrogates to see the baby, delivered borne by the surrogate mother.

Section III: - Proposed New Draft Assisted Reproductive Technology Bill – 2016

In the absence of any law to govern surrogacy, the Indian Medical Council of Medical Research (ICMR) had issued guidelines in 2005, to check the malpractices in India but these guidelines, being non-statutory, were not mandatory, and they were not enforceable in any court of law in India. But the case of *Baby Manji Yamada* Vs *Union of India* had triggered the debate based on surrogacy. Hence the Government of India had introduced the draft ART Regulation Bill. The draft was revised in 2008, 2010 and 2013, based on the suggestions of the Ministry of Law and Justice. According to the Bill, ART in India would be available to all persons, including single persons and foreign duos. However, under the new Medical Visa Regulations, 2012, gay couples and unmarried couples cannot avail surrogacy in India. But the ART Regulation Bill, 2014, restricted surrogacy to infertile Indian married duos only and debarred all foreigners, except overseas citizens of India, People of Indian Origin, Non-Resident Indian and foreigners who had married Indian citizens, with two years of marriage. The draft ART regulation bill was proposed to check escalating surrogacy industry, flourishing in the absence of cohesive Legislation. Mushrooming IVF and ART clinics deliberately advertised for providing

Wombs for Rent. With such a scenario, the Government of India proposed the draft ART Bill 2016, which had banned the commercial surrogacy because it considered it a "rent-a-womb" industry, that exploited impoverished young women. The draft bill is being analyzed critically in Women's Studies Perspective, raising certain fundamental questions based on the Indian constitution, as to how does banning of commercial surrogacy violated and discriminated the rights of woman, as a human?, how banning of surrogacy would`l encourage underground illegal multi crore businesses. Further, it was discussed whether banning of surrogacy would protect women from victimization and vulnerable exploitation. The interview was conducted with various stakeholders including doctors, associations and surrogates, from different cross sections on the Draft Bill 2016. The results of interviews are presented below;

Perceptions and Stand point of the Doctors, Associations and the Participants

As per the New Draft ART Bill 2016, the Government of India banned completely the practice of commercial surrogacy and allowed only altruistic surrogacy. In addition, it proposed some other regulations and new provisions. In order to get the perception and view points of various stakeholders, connected with the surrogacy, interviews were held with Doctors, Associations and the participants of the current research. They are summarized and presented below.

DC1 said, "*As a doctor I would like to appreciate the government for having proposed for such new bill on Surrogacy primarily focusing on banning of commercial surrogacy which I feel, would protect women from the exploitation in the patriarchal and capitalistic context. Men in general force their wives to practice surrogacy whether the women have interest and willingness or not. It is equal to exploitation and the economic support generated has never been used for the benefit of the women who served as surrogate. Women often accepts this for the sake of helping the infertile couples to have a child.*"

It is good that the doctors, who are also part of the stakeholders, could appreciate the provisions of the bill and also banning of commercial surrogacy. It would be a loss for the practicing doctors on infertility, but still the of doctors fraternity recorded that the Bill safeguards women from being exploited by the market and also by their own family members. In fact, in this study, it was quite rare to observe any husband forcing the wife to undergo surrogacy. Rather in several cases, it was the women who surrogated themselves willingly and accepted to serve as surrogates for economic reasons.

DC2 said, "*As a doctor, I accept the banning of commercial surrogacy. I don't want commercial surrogacy to be practiced as it has emerged as a huge business and competitive market. The government provides only legal provisions and often fails to implement the law. The government must strictly implement the law to protect the women from being exploited anymore. There should not be any loop hole to use that as the scope for practicing surrogacy. At the same time the banning of commercial surrogacy will affect the couples who wished to have their biological child when they do not get their close relatives to practice altruistic surrogacy.*

Some women who have all capacity to bear a child but the uterus is weak, would be advised to take a surrogate support. But in case no relatives will be willing to accept the demand that too without any incentive, the possibility of getting a genetically related child will be an issue. They must resort to 'adoption' which they may not be interested. In such cases, the entire banning of commercial surrogacy has turned out to be shutting the door for needy couples to have their own biological child though the technology is very much present.

The doctor added that there are other serious issues: The existing surrogacy provision maintains a complete confidentiality between intended parents and surrogates. It is to protect the child from being not claiming ownership on future. But as per the altruistic surrogacy, the surrogate mother will be known openly and the confidentiality will be lost. This will have serious implications on the parents and the child in future. What will happen if husbands turned negative showing the practice of altruistic surrogacy may try to marry some other women? So banning commercial surrogacy and encouraging only altruistic surrogacy, without mentioning the confidentiality may lead to loss of married life and smooth relationship with the partners. It is quite dangerous."

In extreme cases, surrogacy was viewed or equated with sex work. The pertinent question is: what is the guarantee that the altruistic surrogate mother will not claim or reveal the identity to the child in future as his or her surrogate mother? What kind of bonding or relationship will be maintained?

The DC2 *added that as far as I am concerned. I will only suggest IVF as the best method, because I don't want to get involved into legal issues for the sake of helping the commissioning parents who seek for surrogacy support. Given that the Doctors who practice surrogacy normally take such treatment as commerce and business. It is only a very few doctors but exploitative and business attitude of such a few doctors lead to the banning of commercial surrogacy where such doctors almost treat the surrogate mothers as house arrested for ten months. They strictly prohibit their mobility and contact with their family. All these have added to be the reasons for banning commercial surrogacy.*

From the doctors' view point, the illegitimate act of a few practitioners could lead to non-availability of the facility to the entire lot. Exploitation of any kind must be condemned but when it comes to eliminating an existing facility for the deserving people, need not be encouraged. Every individual must be vigilant about the misuse or exploitation of a provision as it is available to people who are in desperate need. All stakeholders concerned must attack and try to plug any deviation from the law. It is true in the commercial surrogacy context, where the facility was misused. But banning the practice would affect the couples or other stakeholders, who had a hope to have their own baby.

Perceptions of the Surrogate Mothers' Right Protecting Associations

The perception and of the Surrogate Mothers' Right Protecting Associations is presented below,

ASO1 shared, "*The government is just pushing it to the black market.*" *He says the ban on surrogacy for foreigners raises a lot of questions since exporting embryos is not permissible, which means those foreign couples who have their embryos here cannot have them sent back to their countries. If the government wishes to protect poor women from being exploited, it should regulate the sector rather than banning it.* "*The surrogacy bill does not make any provision for the protection of women, assuming that banning commercial surrogacy will protect them.*"

ASO2 shared that, "*When I heard about banning of commercial surrogacy, I really got fed up with because we who are working at ground level, know the hard and ground realities. Before the banning surrogacy, the government should have invited the Association for discussion the proposal, being the Associations working for the rights of Surrogates and well informed about the process and the kind of issues, exploitation etc.*"

It should have been a sort of regulating commercial surrogacy rather than complete ban as the Associations which are genuinely working for protecting the rights of the surrogate mothers, with several safeguarding strategies. The associations' stand point that instead of banning, the government must have gone for strict regulation and vigilance and monitoring, sounds a way of providing a balanced approach rather than completely banning and giving scope for illegal surrogacy practices, which would still be exploitative and unsafe for the surrogate mothers. Because in the current context of banning the commercial surrogacy, in case if a woman practiced and encountered any issue, she cannot even legally seek remedy because, as the commercial surrogacy illegal, became as per the New ART Bill 2016.

ASO1 "*Added further that she spoke on these issues at Delhi in several meetings and surrogacy conferences regarding legalization of surrogacy. But it was never accepted and they are very strong against commercial surrogacy and decided to ban by all means. It was reported that due to the banning of commercial surrogacy which happened suddenly, it had affected both the infertile couples and surrogate mothers who are equally suffering due to loss of such facility which was fulfilling the needs both at the demand and supply side. Also, the govt. was not ready to hear the voice and pain of the doctors, association, infertile women and the surrogates. Most importantly the doctors are not ready to organize talks with the government. The doctor also added that it is also true that 'We the doctors could not talk against the government. If we stand against power of the Government, the doctor's fraternity will suffer. Hence, instead of arguing against the government banning the commercial surrogacy, we just remained calm for our safety and security.*'"

So at the end of day, it is the individual safety and security which mattered and not the people's issue. The doctor's interviews did bring mixed responses and self contradicting also. On the one hand, in the very beginning, it was sold that the doctors were for banning the surrogacy, but later as the interview proceeded, they began saying three significant issues: the secrecy to be maintained; the claim by the surrogate; and the vulnerability of the altruistic surrogate mother, losing her own life partners in extreme cases. Finally, the doctors' voice was not heard not the doctors went against the government, meant that there was core 'self interest' rather than concerns over the issue.

Surrogate Mother's Perception

In addition to the interviews, held with Doctors and Associations, to get their views on the New Draft Surrogacy Bill 2016, further interviews were also held with surrogate mothers, who were part of this study. The view points of the surrogate mothers are produced below.

SM1 said, "*Instead of banning surrogacy it should have been legalized as the permission to commercial surrogacy helped both the surrogates and commissioning parents mutually where the former could get their economic needs met and the latter could get a biological child. She also added that not all the surrogate women practice only for commerce. Those who practice only for money must be discouraged. Those women who suffer with extreme poverty and livelihood needs alone be allowed to practice surrogacy. She did mention the categories of women such as widows, separated, divorced and the women who do not get any support from husbands; the banning of commercial surrogacy would again push the women into poverty and hunger.*"

It is a genuine argument, from the participant perspective, that the surrogate mothers opt for surrogacy out of poverty and hunger and once it is banned, these women and their households would be further pushed to abject. As Pareto Optimality Theory in Economics claims, "no one will be made better off without making others worse off" very much applies here. But the point is that, the market, the exploitation, the capital accumulation attitude and feudal culture being prevalent among the capital owners, do not have any respect for humanistic, ethical and rights of individuals and they indiscriminately exploit the poor. Such activities sometimes affect the entire lot, which is why the ban on surrogacy as per Draft New Surrogacy Bill 2016.

SM2 said, "*I don't know why government banned the commercial surrogacy because I had undergone surrogacy twice because of that now my family is in a better economic position and freed from poverty. Like me more there are women benefited from the commercial surrogacy. But now such women will suffer due to the banning of commercial surrogacy. It will equally affect women who seek for infertility treatment through surrogacy. Such state would exclude these women from the society and label them as infertile, impotent women. Instead of banning, the government must only regulate the process and legalize surrogacy with strict monitoring process.*"

Doctors, associations, and surrogate mothers shared that commercial surrogacy should be legalized because the ban would cause distress for infertile couples and surrogate mothers. This process of the government should not push infertile duos and surrogate mothers to commit a wrong against law, for something as basic as parenting. There is no guarantee that the commissioning parents and surrogates would not secretly engage in illegal surrogate agreement or contract. If the doctors do not agree, they may resort to quacks also in extreme cases. Hence the government must have a second thought, over the provisions of the new draft surrogacy bill 2016, by holding discussions with various stakeholders, who are better informed about the ground realities and the scope for violations.

Community perception

It is very important, to know the opinion of the people, who one among in which will help to understand the ban on surrogacy to the core.

C1 shared "*Surrogacy is not less than a boon for infertile couples. Infertility is a curse in our society. Surrogacy gives such couples a chance to get their own biological child. Some gay couples also got benefited by this. There are a lot of moral and social questions raised on it, like when there are so many homeless children, why can't couples adopt them? In some religions it's equal to sin. Some people say it may increase the number of gay marriages. In my opinion, there should be an ethical and legal guideline for surrogacy. One more important thing, it is becoming a mean of easy money for women in poor countries. Giving their wombs on rent for the sake of a big amount (not for serving the humanity) does not allow me to respect them. It is more like a business for them. That is why surrogacy is banned in some countries.*"

C2 shares, "*First of all, let the women decide if they want to be surrogates. That is the first step in female liberation - let women makes their own choices about what to do with their bodies. Passing laws and bans on what woman can and cannot do is patriarchy in action. If a woman chooses to use her body for labor and profit (the term is called "embodied labor"), this is the choice of the woman. India's banning international surrogacy does not protect Indian woman; it merely paints a media picture that India is respectable to woman. In fact, India is not respectable to woman and this has not to do with surrogacy, but with the fact (just to name a few horrible things) that the Indian government permits killing women (dowry deaths) and forces sterilization on rural Indian woman. Surrogacy, in fact, gives rural woman a chance to make more money than they normally could: becoming a surrogate for 9 months can make an Indian woman the same amount of money as if she worked full time for 3 years. Do not pretend that banning surrogacy for international parents protects woman. All it does is push the industry underground and puts Indian woman in more danger. By making surrogacy only allowed for Indian couples, this helps insure that what bad things happen in India stay in India. India is trying to polish up its international image NOT protect women! This ban avoids coming up with real and safe solutions - which would be insuring that clinics pay the surrogates fully and that the surrogates are cared for well during their contract. India, your government and law processes are shameful.*"

According to them, instead of banning surrogacy, it should be legalized. Each coin has two side and, each and every process has both pros and cons but banning again, will only lead to distress to the deprived women.

Regulating Commercial Surrogacy: Possible Arguments and Way Forward

One of the major aims of the New Draft ART Bill 2016 is to regulate surrogacy where commercial surrogacy had made it a typical market and the intended couples from abroad conveniently utilized medical tourism, with the least cost possible. But a lot of issues cropped up during and after the process. The classic case of Japanese couple who got divorced and both disowned the already committed surrogate child and it led to legal issues. In Gujarat, Anand (2008) through surrogacy, got a female baby for a Japanese duo but by the time of delivery, the duos got separated and they left the baby alone in India. The Japan Government also was not ready to accept the baby because in Japan, commercial surrogacy was not legal. After a long legal battle, the Gujarat high court granted custody of the baby to her grandmother. But, the baby was denied Japanese citizenship since surrogacy was not legalized in the country.

To regulate the commercial surrogacy, the Government of India banned. of hiring the Indian surrogate mother for foreigners. But as per the 2009 Law Commission Report, the artificial reproductive technology treatment industry was Rs. 25,000C industry. However this banning, does not guarantee that surrogacy will not be practiced by illegal means. The intermediates and agencies would definitely identify ways and means, to encourage the underground illegal practices for their well being and adopt different strategies, with enormous implications for both demand and supply sides. If underground black market were allowed, to be the Government has to face multi-crore losses, due to the ban on commercial surrogacy, which have drastic implications for the GDP growth of the country. It is equally true that such ban would lead to multiple exploitations in the society. Owing to the ban, less privileged deprived women will suffer a lot due to multiple exploitations. The banning process restricts women's decision making capacity and it further leads to feminization of poverty. Hence national and regional level consultations with stakeholders, must be organized before passing the bill into Act. Researchers, associations, doctors, surrogate mothers, egg donors, public, women's group and the bureaucrats must be invited for consultation and serious thought must be given before taking any concrete decision on banning or regulating the commercial surrogacy. The feminist concerns and the violation of the rights of women's autonomy, with non-patriarchal viewpoints, must be recognized and respected in the final bill.

To sum up, the New Draft ART Bill, 2016 is no doubt a step forward but in fact the way in which it has been designed and proposed to be implemented, has serious implications Before the evolution of reproductive technologies, infertility was seen as curse for a woman and infertility women were excluded in the society and in a patriarchal society, there is a social norm that those women are not fit to live their life. The patriarchal society always considers women as reproductive object and hence barren women faced multiple abuse in the society such as oppression, suppression, sub ordination and marginalization etc. Banning is not the only solution for each and every problem. The most important thing is that women have internalized the norm that procreation alone gives meaning for the entire life. Under such circumstances, revolution of reproductive technologies was seen as boon for infertile women through surrogacy. In this scenario, the Government of India recently banned commercial surrogacy but banning of commercial surrogacy is not accepted by the infertile couples and the surrogate mothers. Banning has not been fulfilled the requirements of the surrogates and the childless couples. It has to be redrafted, with the incorporation of proper humanitarian benefits for the surrogates since they lose their social image, health and their economic process. Surrogacy has been an unregulated arena and there is immense scope for exploitation of poor women acting as surrogate mothers. It is the single major reason that called for the regulation of the transaction. The intended parents and surrogate mothers expected the Government to come up with the right platform, for the surrogacy arrangement, so that they can work without scope for exploitation. This move of the Government must not push infertile couples and surrogate mothers, to commit a legal wrong for something as basic as parenting. The surrogacy bill must keep all these considerations before finalizing and passing into an Act. The women's freedom and autonomy cannot be given up but at the same time, the exploitation of the marginalized can also be a great concern. It is with such balanced perspective the decision must be made with great caution while regulating surrogacy.

Inferences

From the above analysis, the following inferences were arrived at,

- ✓ The participants came to know about surrogacy through the advisement in the print media and through the intermediaries or agents of various Associations. Capitalism always has a group of agents or brokers as intermediaries, who really fix the terms and conditions for any market, which is true in surrogacy also. It is the brokers, who decide the entry of a woman into surrogate agreement and these surrogate women are at the mercy of these agents.

- ✓ The surrogate mothers in general do not have any knowledge about the surrogacy process nor are they able to get convinced at once. It is the agents or associations, which imparted the knowledge on the surrogacy process.
- ✓ It is poverty and livelihood challenges that women were forced to accept surrogacy contract in this study. Men are expected to be the bread winner and the women's earnings are treated as supplementary. In this context, women serving as surrogacy really brings a boon to the economic status of the household.
- ✓ None of the surrogate mothers informed that they were undergoing surrogacy, except to their parents or husbands. The relatives and others were informed that the surrogate mothers were going to be engaged in the care of the elderly. In the patriarchal context, surrogacy was equated with sex work and it was deemed unethical to bear the child for others. If allowed, it may end up in unnecessary gossip and even breaking of relationship. It is also true that the surrogacy arrangements were kept confidential.
- ✓ The confidentiality was highly maintained regarding the identity of the intended parents as it was perceived that if the surrogate came to know the intended parents, they might claim the right or for the child.
- ✓ The surrogate mothers or members of their families could not read the provisions of various clauses in the agreement written in English. The agreement process, in the capitalist system of surrogate market, just assumed surrogate mothers to be objects and their role was to just bear and hand over the baby delivered. There was no transparency as the copy of the agreement was not given to them. Hence they could not make any claim for any complication post surrogacy.
- ✓ It was revealed that the surrogate amount earned was largely used for paying off the debt, borrowed from the money lenders. It must have been invested on girl children's education and higher education but normally it was not done. Even if it was spent, it was on son's education as the gender perception was male perspective in nature.
- ✓ The unproductive expense such as alcohol consumption, brought about huge gap between income and expenditure and debt. In order to pay back such accumulated debt, which was due to the irresponsible act of men, pushed women to vulnerability. From a feminist perspective, women are always pushed to come to the rescue of households from the financial problems, but such contributions were recognized. Women always volunteer to work, to redeem the family from being pushed to

indebtedness and in extreme cases, for the sake of family, they were even pushed to take up commercial sex work also.

- ✓ The family members were not allowed to meet the surrogate mothers, who were staying at shelter homes. This aggravated the emotional need of the surrogate, though the women asked to be prepared to face the parting with family. Surrogacy contract was accepted for payment and women must also control their emotions, though against nature. Such emotions might affect the psychological and physical health and thereby the baby too. This must be recognized by the intended parents and surrogate must be allowed to meet children and husband.
- ✓ On confirmation of surrogate pregnancy, surrogate mothers were advised not to have their regular and physical activities, except for nature's call, and they not allowed regular food. All concerns were on the well being of the fetus and the baby. As per the *Bonded Labour System (Abolition) Act, 1976 and by Article 21 and Article 23 of Indian Constitution, bonded labour system is abolished, but the* modern technology promotes modern slavery or bondage which is revealed in the surrogacy market. As per the Indian constitution, each and every human being has equal rights. But capitalism has indirectly created reproductive role of women to a modern bonded labour system, with the help of technology. Reproductive labour is an alienated labour, and the law should protect potential surrogates from alienation. Surrogacy should not make a woman to regard herself as a "reproductive vessel." Surrogacy alienates women from their own reproductive capacity and products.
- ✓ The response was mixed, with respect to care and support from intended parents. In general, there was enormous support during pregnancy, by the intended parent, which indicated that the surrogate mothers received nutritious diet, care and other support to safe and healthy delivery. The partners of the surrogate showed enormous and special care during surrogacy, which might be due to two facts. (i) the lump sum the surrogates would be bringing (ii) parting with the family and the recognition of the sacrifice made for the sake of family, as pregnancy is also painful and it amounts to invited risk, pain etc. However, women are still treated as secondary and recognized only when such great efforts are being made by women, which may not be the case when women do not engage in such support systems. The market is also exploitative because it extended extreme care till the work was extracted and least care was shown once the production relation was completed. It is true in this study, that capitalism market used women as objects to be used and thrown away after use. If the women, by choice, were interested to have their own biological child, it may not be an issue. But such choice was due to the social pressure, they opted for ARTs. It is to be underlined that in the traditional

context, when the technology had not developed like today, there were arrangements for the husband to get into second marriage, which was accepted by women also. In such a context, both women were victimized and had to accept the social norm. When such traditions continued, several legal initiatives had to be including Indian Divorce Act. But such trend continues still in rural context where women, who do not get a child or a male child, are ready to identify another woman for the husband to marry. Such regressive customs were challenged by the advent of technology but still women are treated as procreators.

- ✓ The exploitation of the surrogate market cannot be explained as the compensation, for the kind of labour delivered and also the psychological pain undergone by staying away from the family and completely disconnected from contact in certain cases. The surrogates missed their children too. Itvwas disheartening to learn that agreed payment was not paid in certain cases. There was gap between the doctors and the intermediaries in one or two cases where the doctors informed the surrogate the amount paid to the agent but the agent absconded or cheated by paying a partial amount. It is in this context the rights of the surrogate mothers had to be safeguarded. The intended parents must also be connected to the payment and they should see that the agreed payment to surrogate mothers was completely paid.
- ✓ After fertilized embryo transfer (ET) or blastocyst (BT) transfer, for confirmation of surrogacy pregnancy, injection would be given to the surrogates on hip, thigh, vein etc. After pregnancy confirmation, several tablets and injections were given to induce hormonal secretion, which might have serious implications post surrogacy. It is corroborated in other studies also.
- ✓ The Center for Bio Ethics (2013), in their study, found that tablets and injections, used in surrogacy, did have minor and major side effects after surrogacy.
- ✓ Commercial surrogacy should not be banned in India because banning of surrogacy leads to multiple exploitation of women in the society and it would be encourage practice of surrogacy through clandestine ways.

To sum up, the analysis and discussion chapter on surrogacy, clearly revealed that various dynamics were experienced, both in the private and public spheres, by the surrogate mothers. Market really fixes the terms and conditions, and likely to exploited the labour of the surrogate mothers, who actually agreed to the surrogacy contract for economic reasons. ARTs and surrogacy are gendered technologies, with highly specific and differentiated applications, on men's and women's bodies (Konrad 1998). Commercial surrogacy is an honourable profession for women because are cleaning houses and offices, providing child day care, clerking and waitressing more honourable than bearing and giving birth to children? (John, 1993) A few of them had mentioned that commercial surrogacy is equal to sex work. Banning surrogacy will not end the practice and it will simply push it the black market. Instead of banning, right-based, legal framework might provide benefit to everyone.

The reproductive autonomy helps women towards their empowerment. The perspective varies from person to person but if the women want to be economically independent, they may take their decision according to their needs and desire, which are the basic rights of the human being. The responses and experiences of the surrogate mothers did not conform to the extreme cause of exploitation and ill treatment. However, it cannot be completely denied that the surrogate mothers did not face any exploitation. The procreation role of women has been emphasized, where, in certain cases, the surrogate mothers admitted that they had accepted the proposal for the sake of helping the infertile couple. Such perceptions must change. For the sake of support of the technology, it is not necessary that women must have their own biological child because adoption can also be taken as an option. The market must also emphasize this option of adoption in which an orphan will be safeguarded. Family and capitalist patriarch influenced both the demand and supply side of the surrogate market. There must be awareness on adoption, among the barren couples.

Chapter VII

Summary of Findings, Recommendations and Conclusion

This Chapter presents the summary of findings and offers suggestions and policy recommendations. There are two sections, in which the Chapter has been structured. The first section deals with the summary of findings, followed by the section dealing with recommendations, based on the findings.

Section I: - Summary of Findings

The aim of the study was to explore and analyze the rights, issues and challenges faced by the surrogate mothers, the process of surrogacy in which they were involved, acceptance from family and related issues, faced by them in terms of their rights, health, payment, exploitation, caring the family members etc.

Accordingly, the objectives were framed and the data were collected. As already mentioned, the data collection had to be abruptly stopped due to the issues faced by the surrogate mothers in the Association and the co-operation received through the Association, to get the data with additional members, could not be continued. Hence only a limited number of thirty three sample surrogate mothers alone could be included and the data collected were analyzed and discussed. The study proceeded with the following objectives:

Objectives

- ✓ To study the socio economic, demographic characteristics of the surrogate mothers.
- ✓ To identify the motivation, obstacles and facilitative sources to surrogacy.
- ✓ To analyze the process of identification, persuasion and compromises the women made, the nature of market through intermediaries for surrogacy, and the associated issues, the gap between the terms and conditions agreed upon and actually implemented.

- ✓ To find out the issues faced by the surrogate mothers post surrogacy – emotional, psychological, physical health and trauma and strategies to manage the same.
- ✓ To present the perception of various stakeholders like doctors, associations, surrogate mothers, community and present the reflections on the New Draft ART Bill 2016 with the pre and post Bill changes and safeguarding mechanisms.

a. Socio-Economic Background of the Participants

Surrogate mothers reported different kinds of experience, as they were drawn from various socio-economic backgrounds. Irrespective of age groups, educational strata, and occupation, type of marriage, family type, and area of residence, community, religion and standard of living the surrogate mothers faced poverty, indebtedness, lack of livelihood opportunities etc, which were quite common among all. Specific findings, on their profile, are summarized below.

- ✓ The average age of the participants was 31.18 years. Majority of participants (39.4 percent) accepted surrogacy, in the age group of 26 to 30, after having completed their reproductive role in the family, with a minimum of two children.
- ✓ 27.3 percent of the participants were married before the minimum 18 years of age for marriage. In the patriarchal society, parents considered girls as a burden because they had to maintain the so called virginity, safeguard them from various abuses from the society and maintain their morality. It was one of the major reasons for early marriage.
- ✓ Marital status of women revealed that all participants were married. There was no case of unmarried participant in the study, indicating that only married women could undergo surrogacy process. It is a cultural expectation also. It is due to such cultural and legal barrier that no unmarried woman could opt for such a profession. Moreover, in the Indian culture, pre marital pregnancy is never encouraged or recognized and hence even if it is for surrogacy, it is quite difficult to entertain, from the mainstream perspective.
- ✓ Majority of participants (81.8 percent) lived with their husbands and children. Some of them were widows. Widowhood, without any support, might have forced the women to opt for surrogacy.
- ✓ More than half of the participants (57.6 percent) reported that they had their marriage, as per their choice, and remaining 42.4 percent of the participants reported family arranged marriages. It is to be underlined that more than half of the participants had opted for marriage by their choice. Incidentally, such marriages were often not supported by either side of the parents and the couples are forced to start their life afresh, without any financial support from parents. In such a situation, in the

absence of attractive employment with decent wages, poverty of such households 'surrogacy offer' served as a boon and they opted for surrogacy to support the livelihood needs of the family.

- ✓ Around half of the participants (48.5 percent) reported two children. The average number of children was two. It is to be noted that no woman shall act as a surrogate, for more than five successive births in her life, including her own children. Hence it was quite rare to see the sample households, with more than two children except a few.
- ✓ The educational status of the participants revealed that only around six percent of the participants were graduates and diploma-holders. More than one fourth of participants could only sign, followed by an equal number of (24.2 percent) participants were qualified at primary and high school level. 15 percent of the participants had completed their middle school level. Regarding work participation rate, participants in paid employment had been quite low and more than half of the participants were home makers (54.5 percent). Majority of those who were in paid employment (55.5 percent), were found in the informal sector, which reflected the macro trend in India. Whether women were in paid employment or not, the financial access and autonomy of women was poor and had to opt for surrogacy to earn a lump sum, to pay off their debts. Majority of surrogate mothers and their husbands were in the informal sector of employment, which was characterized by exclusion from the organized labour market and also with issues of low wages, home-based work, insecurity of tenure, lack of benefits, poor working conditions, with implications for the health of workers etc. Given this scenario, surrogacy provided an option for enhancing their financial security.
- ✓ Roughly around a quarter of the participants (24.2 percent) were from the middle income group. It was observed that there was a huge monthly deficit, as the income of the households of the participants had been quite low compared to the monthly expenditure. Such monthly deficit often forced the households to resort to borrowing, leading to in accumulated debt, with principal amount and the interest not being paid in time. For fear of attack by money lenders, such poverty ridden households went for another bout of borrowing, to pay the earlier debts and the vicious cycle of borrowing continued. Apparently there was no other source of income, to fully pay off the debt. There were only two options such as migration or surrogacy. In this study, they preferred surrogacy.
- ✓ Adding fuel to the fire, the study revealed that as high as 69.70 percent of the participants' husbands consumed alcohol on a regular basis (every day or nearly three or four times a week). Such facts doubly forced the members of the family, to seek some other employment or the other. Monthly deficit and spending on unproductive consumption

like alcohol, had implications for their livelihood. The woman in the family often came to rescue by way of seeking the support of parents, friends or relatives or go to work, which will bring some advance be it domestic work, construction work, agricultural contract work, companies and factories, mines etc. In this study, it was surrogacy. In general, when women do not get the support of the male members fully, to meet the livelihood needs they often resort to construction work in urban and semi urban regions, and agriculture and allied activities in rural areas.

- ✓ The monthly average alcohol expenses were reported at Rs: 3439/-. It had ranged from Rs. 1000 to Rs. 4000 in the case of 30 percent of the participants' husbands. There were households which reported nearly Rs. 6000 on alcohol consumption, spent by the husbands of the participants. This was one of the key reasons for surrogate mothers to search for some employment. When surrogacy offer came, it was easily picked up.
- ✓ As a consequence of monthly alcohol expenses, ranging from Rs. 1000 to Rs. 6000, take home the earnings of the husbands was quite low and hence majority of the sample households experienced object poverty. As high as 75.8 percent of the sample households had reported that they suffered from huge debt. As already pointed out, such trend continued as a vicious cycle, with the poor households, being reduced to vulnerability. In order to repay the accumulated borrowing, the sample participants had no choice to decline when the surrogate mother offer came. Due to such issues, as high as 93.9 percent of the participants did not have any savings.
- ✓ As high as 87.9 percent of the participants lived in a nuclear family. Living in a nuclear family gave freedom to the couples, to make a decision on surrogacy. Majority of the participants, who opted for surrogacy, could convince their partners and agreed to the terms and conditions for surrogacy. It was mutually helpful to those who lived in joint family, where the children were taken care of by the elders in the family, during the surrogacy period. Hence the participants' decision on surrogacy was decided by the structure of the family.
- ✓ One of the major determinants of the economic background of the family is the size of the family. There is an inverse relationship between size of the family and income. Larger the size of the family, lower will be the distribution of income per head. This study found that as high as 70 percent reported four to six members in their family size. The average size of the family was four, which conformed to the small family norm in India. The size of the family is associated with economic vulnerability and it forced women to accept surrogacy.
- ✓ The religion wise distribution indicated that majority of participants (81.8 percent) were Hindus and others were Christians and Muslims.

Actually Islam does not encourage such practices but the poverty, among Muslim households, drove them to surrogacy.

- ✓ With reference to community wise distribution, majority of the participants belonged to economically and socially challenged communities, with nearly half of them from Scheduled Caste. The rest belonged to backward communities. Only one participant was from the upper caste. It was not surprising that more percentage of historically disadvantaged and marginalized communities, which were also economically backward, opted for surrogacy. Stigma and myth or misconception associated with surrogacy, treating it on par with sex work, was the perception of the middle class. The lower income group subscribed to such a perception cannot afford to maintain but could not sustain this perception due to economic needs. The cultural binding, socio customary norms etc were ignored when poverty and hunger were pressing. In extreme cases, it would lead social crime activities in general, irrespective of men or women. Hence the State and civil society have the duty to keep the people poverty free.
- ✓ As high as 93.9 percent of the participants resided in urban areas, thanks to the urbanization, which had offered support for the people, who were driven out of the rural areas due to the poor or failure of monsoon and agriculture. Moreover, it is quite true that urban participants enjoyed better access to medical services, to practice surrogacy.

Age, income, economic participation, community, religion, region of origin etc did not influence women to become surrogate mothers. The surrogates, drawn from different caste, class, region and religious background, reported that patriarchy was the fundamental factor, coupled with capitalistic mode of production and marketing responsible for surrogacy.

Factors Influencing to Surrogate Motherhood

Several studies have evidenced that the economic vulnerability, in terms of irregular and lowly paid employment, low and absence of income, huge accumulated debt, ever increasing household expenditure due to inflation, lack of alternative livelihood options, poor wage below the minimum wage in agriculture and other non-farm employment, lack of knowledge on regular employment opportunities, structural difficulties towards migration etc, forced the women from such households, to hunt for any livelihood options. In general, these women choose to be domestic maids, construction workers, laborers in garment sector etc. Women did not acquire any skill not had the advantage of and married off in their early teenage. When they were pushed to meet the livelihood needs, due to the above factors, they choose to learn tailoring or any such home based activities or chose to paid employment in the informal sector, which did not require much technical skill. The notion of patrilocal marriages subscribed to the standpoint that the investment on education and skill development for women, was a liability and such investment did not bring any return to the family. Against this background, some women, when an opportunity was open to take them to hire out their own uterus they accepted this offer with lot of persuasion, convincing and self consoling, in order to support the livelihood needs of family or get rid of the huge indebtedness or meet the medical or educational needs.

a. Source of Information

The information about the surrogacy offers was not easily available and accessible to the sample women. It was only through other surrogate women, or egg donors who turned surrogates they also turned surrogates. In fact not all categories of women would opt for surrogacy. It was only in certain regions that surrogate mothers were observed. In other words, surrogacy information was not easily available. But still there were advertisements, given by commissioning parents and the agents were engaged as intermediaries between the commissioning parents and the surrogate mothers. When the surrogate mothers, in the study, were asked about the source of information, the responses were:

- ✓ More than half of the participants (51.5 percent) informed that the intermediaries and agents were the sources of information.
- ✓ A little higher than one fifth of the participants (21.2 percent) could access the information, through print and television media.
- ✓ The surrogate mothers, who had served as surrogates earlier, egg donors etc, formed another source of information.

b. Motivating Factors

- ✓ As already elaborated, majority of surrogate mothers chose to be surrogates for economic reasons. It was one of the key reasons for accepting surrogacy because they had to repay accumulated debts in the family.
- ✓ The economic profile of the participants revealed that they were engaged in low paying, seasonal and informal work. This landed them in financial difficulties. The remuneration for surrogacy, the amount, the nature of payment (lump sum), and the time span over which the amount was received, were the vital reason, for becoming surrogates.
- ✓ Education and the future of their children etc, were the common reasons shared almost all the participants. The participants insisted that they underwent surrogacy for the sake of their children's future and welfare.
- ✓ Only a few of the participants (6.1 Percent) were motivated to practice surrogacy for hedonistic reason like service, self gratifying, support to the suffering infertile couple etc.

- ✓ Majority of participants convinced themselves, externally and internally, to undergo surrogacy, due to threat of huge debt and attractive payment towards surrogacy.
- ✓ The participants practiced surrogacy only for the well being of the husband and children and not for their own well being, which was not a surprise in the given socialized patriarchal contexts.
- ✓ Regarding the factors to choose to become surrogates, though majority of the participants (60.6 percent) did opt to such service for economic reasons, there were also others, for non economic factors but it was only for just one or two participants.

c. Dilemmas and Confusions

- ✓ Majority of participants' (63.63 percent) attitudes were positive towards surrogacy when they first heard about it. However, around 18.18 percent of participants reported negative perceptions and attitudes. Some of them experienced confusion, regarding surrogacy, as to whether to opt and if opted, what would be the social, physical, health and other consequences. However, physicians cleared their confusion, doubts etc, and they were also emotionally convinced by the Associations and intermediates, on the grounds that it was a social service, which gave a child to the childless couple, whose emotional and psychological satisfaction could never be measured. Regarding the participants' doubts on medication or the process of pregnancy for the surrogacy, were cleared and participants shared this with the Researcher. It is true that medical process cannot be explained fully to make them understand and also the complications, if any, during the surrogacy process, may not be informed for the reasons that they might get panicky and give up surrogacy.

d. Persuading the Life Partner/Parents

- ✓ In several cases, self convincing of the surrogate mothers takes place with such mediators, doctors etc, but convincing the family members was a herculean task. Surrogacy was deemed to be equivalent to sex work or how will a woman, after delivering a child, depart immediately after the birth. It is revealed that some participants' husbands (21.21 percent) had strong doubts about surrogacy process because they assumed that surrogates would be required to have sexual relations, which was tantamount to sex work and hence they had their own reservation to accept their wives to serve as surrogate.
- ✓ Some of the participants were let down by their husbands, due to their wrong notion and understanding of surrogacy process.

- ✓ There were mixed kinds of convincing and persuasion needed to convince the husbands, as it ranged from persons who completely mistook it as sexual intercourse for surrogacy to health concerns of their wives. The surrogate mothers had to take a lot of initiative to convince and persuade, through doctors, agents and the self. The justification given for self convincing was applied to convince the husband, which was none other than economic vulnerability in terms of accumulated debts.

e. Enrollment Process

The identification of surrogate mothers, convincing them to accept the offer, later convincing the family members etc, involved a lengthy process. Sometimes, the family members may not get convinced even after adequate persuasion and surrogacy procedure was given up.

- ✓ Majority of participants (72.7 percent) were enrolled through intermediaries, association agents and advertisements, into surrogacy contract.
- ✓ The agents did not directly approach the participants, to persuade them to undergo surrogacy. First they approached them for egg donation and later they contacted the associations, for convincing them to accept surrogacy. In other words process was not directly to surrogacy but through egg donation. Egg donation has been quite common and girls studying in colleges also entered the egg donation market, to make up their college fees. Doctors identify eligible women, during the process of egg donation, persuade them to opt for surrogacy by offering counseling and convincing them about the medical process and its harmlessness.
- ✓ Nearly half of the participants (45.5 percent) practiced surrogacy, only after donating their egg once or twice.
- ✓ Some participants (6.1 percent) directly contacted the intended parents or physician and 18.2 percent participants were referred to by peer friends and family members, to practice surrogacy.
- ✓ During the egg donation, participants' uterus level was examined by the physician. Later those participants were directly contacted by the Associations, for surrogacy process while the physicians informed the association about the eligibility and potential for surrogate motherhood. Hence all the stakeholders were vigorously involved in convincing the women into surrogacy.

f. Medical Counseling and Examination

Medical examination was done for certifying them to be fit for surrogacy as per the norms of the ART Act and Surrogacy rules. The participants reported that the medical examination was a painful process, more than bearing the child. But it was mandatory to undergo all the medical examination, necessary for the surrogacy. It was revealed that the nature of contact, between surrogate mother and clinic, was purely through agents and associations.

- ✓ Participants, who were lactating mothers, were not admitted by the physician for practicing surrogacy, as per the surrogacy norms, because the lactating mothers cannot become surrogates.
- ✓ Before entering into surrogacy contract, participants' health was thoroughly checked by several tests like ultra sound, obstetrics, blood and urine test. Tuberculosis (TB), Venereal Disease (VD) and HIV-AIDS tests also were done by the physician.
- ✓ Before entering into the surrogacy process, majority of participants had been in the profession of egg donation.
- ✓ A few of the participants were advised by physician, to take some tablets, due to the negative blood group and to avoid complication, during and after delivery.

Thus medical fitness and related tests involved a lengthy process and painful medical examination, which had to be undergone, to become surrogate mothers.

g. Abuse Faced by Participants

- ✓ In the beginning, convincing the women to serve as a surrogate used to be a challenge. But once agreed, one had to undergo all sorts of pains and pleasures. In this study, the surrogate mothers expressed mixed responses. Majority of participants (90.9 percent) had a very comfortable time during the surrogacy period, in terms of care, nutrition, emotional support, meeting the needs at home etc and hence those women were willing to offer themselves to surrogacy again. But in some cases, the participants (6.1 percent) had reported some kind of abuses faced during the surrogacy process.
- ✓ Surrogate mothers, who stayed in the custody of commissioning parents or their relatives, experienced severe abuses, in terms verbal abuse, humiliation, compelling to perform all the household duties etc. It was evident, as per the medical treatment that the surrogate mothers must remain in complete rest during their pregnancy. But still it was violated. It was culturally expected that only when the mother was active and involved in all household duties, the delivery would be smooth and normal. Perhaps such customary beliefs would have forced the commissioning parents and their families, relatives to make the surrogate mothers to perform all duties. In certain cases, the surrogate mothers were forced to take medicines or food, though the surrogate mothers were not comfortable or were throwing up and suffered other

such sickness. In such situations, women were humiliated and abused and the principles of market started working where the surrogate mothers, who agreed to everything, must oblige for which they were. A kind of slavery treatment attitude was adopted and, in certain cases, surrogate mothers had felt sorry for accepting surrogacy offer.

h. Feelings Experienced by Participants

Emotional support was necessary for the smooth delivery and health of the mother and the baby. In certain cases of commissioning parents, the surrogate mothers were in the custody of the relatives, who had asked the surrogate to perform household duties, forced to consume medicines, food etc, though the surrogate mothers were not comfortable, and refusing to oblige, resulted in abuse and humiliation. But there were participants, who experienced comfortable stay and also enjoyed emotional support, meeting their family members and allowed to stay in their own families of the surrogate etc. In other words, the comforts varied across the place of the stay of the surrogate mothers.

- ✓ Those surrogate mothers, who were in the institutional custody, received support from various sides, including the doctors, the family members, mediators and others and they experienced comfortable stay. They went to the extent of saying that it was several times better than their own pregnancy in which women were forced to work in the labor market, do all the household duties, till the advanced stage of pregnancy.

i. Issues Faced by Participants

- ✓ The surrogate mothers preferred normal delivery and they were also assured so while entering into the contract. But practically speaking, all surrogate deliveries were invariably cesarean. It was apprehended that if it was a normal delivery, some untoward incident may occur, causing damage to the lives of the child and mother. Hence medically speaking, the surrogate deliveries were generally cesarean.
- ✓ In a few cases, particularly those who had gone through the brokers, failed to get the full payment. In certain cases, where there was direct contact between the surrogate mother and commissioning parents, the payment agreed was not fully paid. In such cases, Associations came to their rescue, to get the payment as the Associations were actually committed to the surrogate mothers. Several surrogate mothers had to struggle to get the payment as agreed.

j. Forced Factors

- ✓ Almost all the surrogate mothers interviewed, explained that they chose to accept the surrogate contract only to pay back the huge debt. Once the amount was paid, they would be relieved from the past debt. Otherwise, they may resort to continuous borrowing and the vicious cycle would continue since poverty and livelihood challenges were part of their lives. Thus to get rid of the debts, the surrogate mothers had opted for surrogacy. But the very objective was nullified, if the payment was not made in time.

- ✓ All the sample surrogate mothers were married. Majority of marriage was by choice and only a few had been arranged marriages. It is inferred that the lower income group couples, who had gone for marriage by choice, did not have any support, from either of their parents. Getting regular employment and income were real challenges, which might be the reason for pushing them to surrogacy.
- ✓ It was quite interesting to find that only one surrogate mother, entered into a contract, where her husband wanted her to become pregnant for the second time, as they had only one child. Hence they chose to opt for surrogacy.
- ✓ The income of the male members, in majority of the sample households was spent on alcohol consumption, which drove the households to poverty. Hence these women were forced to take up some job or the other, and surrogacy was an attractive option.
- ✓ None of the surrogate mothers, in the sample, was introduced to the commissioning parents, prior to surrogacy.Majority of surrogate mothers' attitudes, towards surrogacy, was positive at the first instant, when they first heard about it.

k. Surrogacy Agreement

Since surrogacy is a complicated and risk bearing task, there are strict legal formalities to be adhered. The study revealed the following:

- ✓ Written contracts were made for the surrogacy and the surrogate participants did not get a copy of the contract.
- ✓ Surrogacy agreement was generally written in English. Some contracts were written both in Tamil and English.
- ✓ In majority of the surrogate contracts, husbands of the participants signed the contract.
- ✓ Majority of participants did not have any knowledge of the content of the signed contract.

Women, who were the subject, had to undergo the entire surrogacy process, which was painful, without being informed about the terms and conditions of the contract. This indicated the extreme form of subordination, keeping women in the dark without information and treating them like an agent to deliver the baby, to harvest huge money. In the matter of economic participation and support also, women were not placed on equal footing with them. It is assumed that women were supposed to undergo the process to make the money and it was not necessary that the very subject must be informed about the content of the agreement. Such state continues in several other fields, where women are just treated as dummy and ignorant. This ignorance was an excuse to further exclude them from the process of decision making and the subjugation continues. Feminists attack thus state of treating women as passive partners and demand recognition of women's 'Agency':

- ✓ A few participants were aware that commissioning parents had signed the agreement and some participants' parents, who were widows or divorced, also signed the agreement.
- ✓ Majority participants' were not aware of the medical terms and issues, addressed in the agreement, except the compensation. This was again due to fact that the participants were medically illiterate. It is the duty of the medical professionals to actually inform, in simple terms, the medical treatment so that the participants would have some idea of what medicine was given, the purpose, the nature of support, side effect etc. But in general, the medical fraternity failed to do so.
- ✓ Majority of clauses, contained in the agreement, were informed to participants' husbands verbally. In the case of illiterate surrogate mothers, someone must be appointed to fully read the agreement and explain them in vernacular language, to bridge the gap for the surrogate mothers.
- ✓ It is every one's right to understand the terms and conditions, the nature of treatment, the medical and treatment process and the surrogate mothers, who undergo the physical pain to deliver the baby must know the contents of the contract.

l. Coping Strategies adopted at Shelter Home

In this study, except 27.3 percent of participants, others had their accommodation at the shelter homes or hospital. In such cases, the contact with family was not entertained. The surrogate mothers, though acted as surrogate for money, were still emotionally connected to the baby. Hence there must be a coping strategy, to make their stay comfortable so that the health of the baby and mother would be ensured. The coping strategies adopted, revealed the following:

- ✓ Majority of participants (54.5 percent) maintained regular contact with their family members, during pregnancy, to avoid their loneliness.
- ✓ 42.4 percent were endowed with facilities such as television, bible and books, prayer space, music etc, which helped to provide entertainment and diversion.

- ✓ Some participants (3.1 percent) prepared their minds to stay aloof, till the baby was born and half of the participants stayed with their husbands and children.
- ✓ Those who stayed with the family and those who were allowed to meet their husbands, faced less vulnerability and emotional issues than those who preferred staying aloof.

m. Mental Health Consequences

Surrogate mothers did face serious mental health issues post surrogacy. It might be primarily due to staying away from family, missing the children, missing to attend the family functions, festivals and above all, guilty of failing to inform the correct reasons for staying away from relatives etc. It might be the accumulated feelings of guilt that they were forced to undergo surrogacy for economic reasons. Some participants married by choice, where the support from family members was missing and there was no one to take care of their children. All these would have pushed the sample participants, to a feeling of psychological and emotional stress, which might develop into depression and the mental health issues. It was revealed in the study that each and every participant faced different kinds of mental health problems such as frightened of all noises, depression, anxiety, unable to eat, sleeplessness, low self esteem, felt like crying, mental restlessness etc. The other reason, for the mental health issues, was attributed to the post surrogacy and departing from the surrogacy child. The nature and kind of mental health issues, faced by the participants, are produced below.

- ✓ Majority of participants (21.2 percent) reported moderate level of mental health consequences, such as sleep related problems and unable to eat, especially after surrendering the child.
- ✓ Around one fifth of the participants (18.2 percent) faced low self esteem and mental restlessness, after handing over the child to the intended parents.

The other general findings, based on the empirical and qualitative analysis, are given below.

- ✓ None from the family of the participant forced her to undergo surrogacy as a commercial activity but only for the well being of the family, particularly to meet the economic needs.
- ✓ None of the participants was aware of the decision, regarding abnormality found in foetus, as the physicians never revealed them to the surrogate mothers.
- ✓ It was found that in case the intended parents refused to accept an abnormal child, around 57.57 percent of participants expressed their willingness to retain the child with them.

- ✓ It was also found that due to some unexpected reasons, if the surrogacy pregnancy was terminated, as per the agreement for the well being of surrogates, intermediaries may claim minimum amount from the intended parents.
- ✓ It was also found that lactating mothers were not permitted to enter into the surrogacy process.
- ✓ The average compensation, for the surrogate mothers, was Rs.2 lakhs. But majority of participants (36.4 percent) received the surrogacy compensation, as less than Rs. 2 lakhs. 36.4 percent of participants were able to receive Rs.2 lakhs to Rs.3 lakhs. In case of twins, participants received Rs. 4 lakhs as compensation, which was really a blessing for the commissioning parents who expected one child but got two children. In fact the commissioning parents were generally asked by the Doctors, whether they needed twins and accordingly, the surrogacy procedure was followed.
- ✓ A few of the participants (21.2 percent) received Rs.3 lakhs to Rs.4 lakhs as compensation, which was not the case with others. Several others could not get such huge amount and even in certain cases even the agreed Rs. 2 lakhs was not paid in time. It was the fault of intermediaries and the doctors or the commissioning parents, in majority of cases. The generous commissioning parents would extend support to surrogate mothers post surrogacy, to share the joy of getting their own biological child, but it did not happen with many other participants.
- ✓ It was found that before the emergence of surrogate mother rights association, the compensation was fixed by intermediaries and doctors. But those intermediates did not disclose the fixed amount to the participants. Participants were cheated by the intermediates. But after the appearance of rights association, the compensation was fixed by the doctors and associations. The fixed amount was uniform and it was fixed for a single child as Rs: 3 lakhs, and for twins, the fixed amount was Rs: 4 lakhs.
- ✓ Majority (54.5 percent) of participants received compensation as lump sum (single payment), which was really a blessing as the very objective of taking up surrogacy is to get the lump sum, to pay off their past debts. It is good that more than half the participants could get such compensation in the form of lump sum but the fact remains that a little less than half of the participants did not get it in lump sum.
- ✓ More than half of the participants (54.5 percent) did not make contact with the commissioning parents nor attempted to, as per the agreement signed. It is good that the other 45.5 percent of the participants could contact and reveal their identity.

- ✓ As high as 72.7 percent of the participants and their husbands had experienced harmonious relationship, at the initial stage of surrogacy. Around 9.1 percent participants reported major conflict, which might be due the kind of stress caused by child abuse in the patriarchal context. Hence, surrogate mothers must ensure, in the beginning itself husbands adapt themselves to the absence of the women.
- ✓ 42.4 percent of the participants stayed in a rented home, which was arranged by Associations, followed by 30.3 percent of the participants who stayed in the hospital and 27.3 percent of the participants, stayed in their own house. The third category did not any difficulties, like mental health issues associated with their separation from family. Similarly, participants, who lived in rental houses, were allowed to see their family members. But the others, particularly those who stayed in hospital, had to go through serious restrictions and hardships, impacting taking their health condition.
- ✓ Very few of the participants (6. 2 percent) stayed with the intended parents and their relatives, in rented homes. It was in the relatives' places, that some participants reported uncomfortable stay, caused by being to forced to perform household works, etc.
- ✓ More than half (57.6 percent) of the participants had contact with the hospital, through the agents or mediators or association, followed by 18.2 percent of the participants, who were approached by the clinic directly.
- ✓ An equal number (45.5 percent) of participants had experienced joy as well as fear, before the surrogacy pregnancy.
- ✓ More than half of the participants (66.7 percent) reported that regarding the date and time of delivery and handing over the child to intended parents, it was decided by the physician.
- ✓ More than half of the participants (57.6 percent) opined that after delivery, they were willing to take the child, if the intended parents refused to accept the child. But the issue was whether the payment would be made as agreed, if the child was not taken by the intended parents, where the very objective of getting lump sum to pay the debt or meet other expenses, would be lost.
- ✓ More than half of the participants (63.6 percent) shared that there was no problem in their family life while practicing surrogacy. But a few of them faced some unhappy consequences in their personal life, which may not be a surprise, given that some of the participants did not reveal the actual reason for the absence and in certain cases, the participants had to convince and persuade the men but these arguments were not fully accepted by all members of the family.
- ✓ As high as 84.8 percent of the participants reported that they used the surrogate money to repay their loans and they were relieved completely

from the money lender, which was a great relief. The remaining participants used the surrogate compensation, investing on future needs of the children and medical treatment of the family members.

- ✓ More than half of the participants (57.5 percent) reported that after surrogacy delivery, in their surrogacy period of experience, the most pressing and moving moments were the point of leaving the surrogate child, which moved them to break down and suffer in silence.
- ✓ A few of them (9.1 percent) reported that having agreed for a payment, when it was not paid as agreed, that was the worst moment for them to suffer. Around 6.1 percent had reported that they faced mental and psychological pressure, after leaving the child.
- ✓ Nearly half of the participants (45.5 percent) missed the proximity of their children while practicing surrogacy. A few of them shared that while practicing surrogacy, they felt like they were under house arrest. Such feelings were painful to surrogate mothers. Feminists would like to know whether these emotions were translated into health issues. Surrogates also feel guilty about not bring with children because in a patriarchy based society, women do not completely give up or share their responsibilities with other members of the family including child care. It is mandatory that the mother should take care of children, irrespective of the age of the children, whether they are kids or grown up. The perspective that father and other members of the family can also take care of children, in addition to the mothers, has not been mainstreamed. In such a context, it is not a surprise that women felt guilty when not spending time with the children. This is applicable to the employed women, who go out for work. The irony is that the surrogate mothers do agree to surrogacy offer only for the sake of the family and the children. So either way, it was to take care of the children but being mothers, they felt that their physical proximity with children was missing and hence it was the worst part of their surrogacy period.
- ✓ As high as 90.9 percent of the sample participants complained that they were not allowed to follow their routine work. It was further reported that they had missed some of the routine work like visiting temples and the daily routine household work. As far as women were concerned, irrespective of religion, offering prayers by visiting places of worship has been part of women's life.
- ✓ Around 36.7 percent reported that till surrogate child delivery, the participants felt a feeling of disowning, felt like being tied up with a heavy agreement, no liberty to be on their own except to deliver the baby well. This was an extreme form of emotion, where the surrogate women, though agreed to serve as surrogate mothers due to the issue of economic needs, still as days went by, they felt disgusted, from resulting a sort of disowning the child immediately after the delivery.

- ✓ While practicing surrogacy, participants faced some issues in their family like mother missing their children for a long time, kind of behavioural change in children and wrong notion about surrogacy in the society etc.

There have been several positive experiences, that the surrogate mothers had experienced which are produced below.

- ✓ Majority of participants (68.2 percent) experienced comfort in the shelter home such as excellent personal care treatment, regular nutritious foods, immediate health care facilities, and special treatment, which they may have never experienced in their personal life.
- ✓ Some participants (31.8 percent) enjoyed their freedom from household responsibility, complete relief from day today chores, health assistants' concern, support, love and affection, sophisticated and clean environment, provision to continue hobbies, worship and watching television, care from physician, intended parents and peers.
- ✓ Nearly 81.8 percent of the participants did not face any issues in the shelter home. A very few participants faced some issues in the shelter home, such as gossiping among the inmates, finding fault with any deviations from the set norms, etc.
- ✓ Nearly half of the participants (48.5 percent) faced psychological and emotional issues, due to separation from home for ten months and the associated mental stress and pain, wrong perception about surrogacy and equating it to sex work and exclusion or abuse by in-laws and relatives etc.
- ✓ More than four fifths of the participants (81.81 percent) opined that they did not have any idea about the background of the intended parents such as nativity, religion, caste etc.
- ✓ As high as 78.78 percent of the participants informed that they did not get any information about the number of fertilized eggs, placed in their uterus.
- ✓ Nearly half of the participants (45.45 percent) felt that compared to their own pregnancy, surrogacy pregnancy was quite comfortable and satisfactory.
- ✓ Nearly 87.87 percent had no idea about the name and purpose of the tablet and injection, used during the surrogacy pregnancy.
- ✓ Two thirds of the participants felt that if their health could permit, they would make a choice to practice surrogacy, for the second time, to help a childless duo.
- ✓ 54.54 percent of the participants were not aware of the gender of the child, delivered by them, through surrogacy.
- ✓ 51.51 percent of the Surrogacy pregnancy was confirmed in the second attempt only. When the first pregnancy failed, the doctors took the

participants for the second attempt, with the permission of the intended parents, which was not informed to the surrogate mothers.

- ✓ More than half of the participants (51.5 percent) reported that after delivery, they wanted to see the child and meet the intended parents to thank them in person and know the growth of the child but there was no such arrangement.

Suggestions

On the basis of the findings of the study, the following suggestions are offered, to safeguard the health of the surrogates. Based on the findings of the research, the following suggestions have been enumerated.

- ✓ Having thoroughly revealed in this study that surrogate mothers had opted for surrogacy only for economic reasons, it is necessary on the part of State, to provide or include such women with special provision, to get government support in terms of employment, in the event of banning the commercial surrogacy.
- ✓ Keeping the surrogate mothers in the dark without information about the medicine cannot be continued as it is their right to know what the medicine they are administered.
- ✓ The inter-caste marriages are encouraged and rewarded. Since majority of the surrogate mothers made their own choice of marriage, especially inter caste marriages, with monetary support could be extended. Regular employment, at least for one of the members, either in the private or public or NGO sector, could be provided.
- ✓ Time and again it has been evidenced that poverty of the lower income group has been primarily due to male members spending on alcohol. TASMAC justifies the sale of alcohol by the Tamil Nadu State, for reasons of revenue and to prevent from using illicit arrack. There is need to regulate the prohibition in phases, as alcohol consumption takes away considerable income and the burden often falls on the women to manage and adds to the drudgery of women. Hence there is need to introduce probation to save the lives of the poor, including the surrogacy practiced women households.
- ✓ There must be arrangements, at least over phone or video chat, to make the intended and surrogate mothers to interact, which will address the surrogate mothers' desire to meet the intended parents.
- ✓ The surrogacy contract must be handed over to the surrogate mothers on ethical grounds, to ensure the transaction to be transparent.
- ✓ The surrogacy agreement must be in the vernacular language and in a readable form so that surrogate mothers could understand the terms and conditions completely.

- ✓ The surrogate mothers must be informed and made aware of the members who had signed on her surrogacy agreement.
- ✓ Surrogate mothers must be informed about the medical process, the nature of treatment and the nature of medicines, the purpose for which it is being given, etc to make them understand the treatment, which is ethically required also.
- ✓ Surrogate mothers must have the option to visit the family members or permitted to meet their family, as per their choice, with all medical caution.
- ✓ There must be continuous counselling arrangements, to the surrogate mothers, during the surrogacy period, to address their needs and help their coping strategies. Such counselling would provide peace of mind and indirectly would contribute to the health of the baby and the mother. It will prevent them from falling into the mental health issues, depression, psychological pain etc.
- ✓ When there are any complications, the surrogate mothers must be informed with all caution, to prepare to face any failure of the pregnancy or other complications so that the surrogate could face it with confidence.
- ✓ Since the payment differed among surrogate mothers, from Rs. 2 lakhs to Rs. 4 lakhs, it is necessary to regulate uniform fixation of the compensation for the surrogacy. Such regulation would help to reduce the exploitation of the poor by unethical agents.
- ✓ There should be an inbuilt and transparent system, to get the agreed payment made in time, without any leakage, as it was reported that there was differential compensation and also some of the participants did not get the fixed payment.
- ✓ The agreed amount if paid in one go, will help the participants to repay the debt. Surrogates must be sensitised to enforce timely payment through legal means.
- ✓ As per the agreement signed, the surrogate parents can be permitted to meet the intended parents. Any denial is a violation of the rights of the surrogate agreement.
- ✓ The family members and the children must be counselled to accept the consequences of surrogacy. Family members including the children, might be informed that surrogacy is a service.
- ✓ Irrespective of the nature of stay, be it at shelter home, or hostel or residence of the intended parents, the treatment of the surrogate mothers, during pregnancy, must be made uniform. Providing poor care or forcing to eat whatever the physical condition or to undertake household activities must be monitored.

- ✓ Whatever the sources induced the surrogate to undergo surrogacy, be it agent, mediator or anybody, the surrogate mothers must be given full information regarding and the medical process.
- ✓ The date and time of delivery and the time of handing over the child to intended parents, may be informed to the surrogate mothers also.
- ✓ In case the intended parents for some reasons or the other, were not willing to accept the surrogate child, the recovery of the amount paid, should not be made.
- ✓ The family members must accept the surrogate mothers with warmth at the end of the surrogacy.
- ✓ There must be arrangements, for a smooth transition after delivery, without making the surrogate mother feeling empty, and deserted, as the surrogate mother had performed such a noble service and she deserves to be given a parting with gratitude. The intended parents must understand and it is their duty to thank the surrogate mother, with the promise to support her post surrogacy health.
- ✓ The surrogate mothers, during their surrogacy period, may be given opportunity to visit places of worship, which may help them to be peaceful and in such places, the surrogate mothers may be given an opportunity, to meet their family members, children and husband, which would facilitate them to view surrogacy as part of their routine. However, if the women were to be under medical treatment, where there should not be mobile, such arrangements need not be encouraged.
- ✓ In exceptional cases, the surrogate mothers may be permitted to be in relationship with the child born through surrogacy, when she turns mentally affected due to separation from the child. Given the communication and easy connection, the foreign intended parents may have video chat with the surrogate mothers occasionally, by providing a a smart phone to facilitate the video chat.

Recommendations

On the basis of the findings and understanding the intensity and contribution of commercial surrogacy and scope for exploitation, it is pertinent to make the following recommendations, which may be implemented to eliminate exploitation in the society. To ensure exploitation-free homes and society, the State, family and society have to play a major role in re-socialization that gives equal space and opportunity for women. Both preventive and protective measures are suggested, to ensure exploitation free surrogacy process. The first part of the recommendations, already incorporated in the new assisted reproductive technologies draft bill-2016, are produced below,

Society

- ✓ Infertility is a public issue and therefore, women, who undergo surrogacy, should not be underestimated, discriminated and devalued by the society for their service and sacrifice, irrespective of class, caste and gender.
- ✓ In the case of marriage by choice, parents of both the bride and the groom deserted them and hence they were pushed to surrogacy. It is advocated that the parents should accept the choice based marriage by the couples and extend support to avert the vulnerable situation.
- ✓ Alcoholic consumption by the male members, keeps the household in economic backwardness, which must be reduced through phased moves towards prohibition.

Association

- ✓ Associations have been formed for the welfare of the society. At present, some of the Associations work only for their own economic benefits and they go commercial. Hence associations need to work for social benefits rather than driven by business motive.
- ✓ Before commencing the surrogacy procedure, counselors should give counseling to the surrogates, to help them to cope up with the procedures mentally.
- ✓ The compensation should be directly transferred to the account of the surrogates other than family members.
- ✓ Based on the request of the surrogates, the compensation should be furnished.

Physician

- ✓ Doctors need to concentrate not only on the child but also on the health of the surrogate and should give non discriminatory treatment.
- ✓ Physician should not work for the benefit of intended parents alone. They must consider health of the surrogate mothers also and should not take them for granted.
- ✓ Before stating the procedure, they must inform the possible health complications to the surrogate. Surrogate mother should not undergo more than three trials and it has to be strictly followed.
- ✓ Sex determination should not be attempted and it has to be strictly avoided. Physician should give priority to post surrogacy health issues also and the physician should follow the ART regulation strictly.

Surrogate Mothers

- ✓ Before signing the agreement, the surrogate must read the agreement completely. Surrogate mothers must try to get a copy of the agreement, for future use, in cases of dispute.
- ✓ Surrogate mothers need to abstain from attempting to contact the intended parents and the child, as per the agreement, to avoid complications and feeling mentally disturbed.

Commissioning Parents

- ✓ Intended parents should treat surrogate mother as a human being and if surrogate mothers wish to see the child, they must be magnanimous enough to provide the opportunity.
- ✓ Intended parents must see that the payment agreed was paid to the surrogate.

Government

- ✓ To avoid exploitation, banning cannot be a solution for commercial surrogacy. Instead the Government may need to arrange debate and discussion with researchers, surrogates, associations, intended parents, physicians, and public representatives and based on the deliberations, some measures should be provided to regularize the commercial surrogacy.
- ✓ There should be a platform to explore economic, social and health exploitation of surrogate. There is a need to include exploitation of surrogate under criminal section of personal law. It should be formally informed to safeguard institutions.
- ✓ In India, there are guidelines but there is no specific law. Hence the Government should regulate the surrogacy, to avoid the medical exploitation of women.
- ✓ National statistics, on the outcome of ART techniques in India, may have to be compiled and made available to the public to make the people to understand the medical process and the role of ARTs, it types etc, in the surrogacy process.
- ✓ Appropriate guidelines should be put in place, by the regulatory body, to govern the number of embryos to be transferred, in any one treatment cycle and when to transfer embryos.
- ✓ Surrogate should be adequately informed of the risks involved, the potential benefits that may be obtained, and the possibility of success in their particular situation. Qualified professionals must explain to the surrogate, the complex medical and scientific ramifications of different treatment approaches, in verbal and written form.

- ✓ Best practice infertility treatment guidelines should be developed, for general practitioners and gynecologists, working outside specialist clinics. These guidelines need to be reviewed on a regular basis.
- ✓ Equality before law must be ensured so that the services will be available without discrimination on the grounds of gender, marital status or sexual orientation, subject to consideration of the best interests of any children that may be born.
- ✓ If intended parents reject the child, the Government must take the necessary action to take care of the child and make the child's future safe.
- ✓ Donors may be prohibited from accessing the identity of the children, born through the use of their gametes or embryos.
- ✓ Women, who decide to participate as surrogate mothers, must be entitled to receive reimbursement of expenses, directly from the intended parents or doctors and must avoid intermediaries.
- ✓ Preconception sex selection may be permitted only for the reliable prevention of serious sex linked genetic disorders but not for social reasons.
- ✓ Adoption rules and regulation must be revised, to enable easy adoption, which will partially address the issue of infertility.
- ✓ Before making a choice to become a surrogate mother, they must know the objective and need for surrogacy with all clarity.
- ✓ Government must frame procedures, to standardize the pattern of payment, for practicing surrogacy.
- ✓ Without the permission of the surrogate, nobody should share the details of surrogate and it should be maintained as confidential. In the research also several doctors were approached to provide the details but it was not given for reasons of confidentiality. Hence with the informed consent, the sample participants were personally met and interviewed.
- ✓ A copy of contract should be provided to the surrogate mother, to maintain transparency and proceed legally, in case of dispute.
- ✓ A separate department to be established for ARTs, especially for surrogacy, under the Ministry of Women and Child Welfare Development, to moniter and regulate the benefit and welfare of surrogate mothers and hence registration must be made mandatory.
- ✓ The contract copy ought to be in a stamp paper, to help in legal claim in future, if needed.
- ✓ The contract should be in vernacular as well as in English to facilitate both surrogate and intended parents, to understand the conditions.
- ✓ The surrogacy contract must be signed, in the presence of the surrogate mothers and the intended parent and also in the presence of an advocate, to avoid any legal conflicts in future.

- ✓ In case surrogate procreates twins, she must be paid double the payment agreed upon.
- ✓ Since the baby was delivered by the Indian mother, the Government can consider according Indian citizenship to the surrogacy baby.
- ✓ The Government must arrange for an inbuilt monitoring mechanism to monitor surrogacy clinics, for the well being of the surrogate mothers and intended parents.
- ✓ As per the Indian Penal Code (IPC), each and every citizen has equal rights and hence so if a woman wishes to practice surrogacy, the law must permit the woman to practice surrogacy.
- ✓ The Government must periodically and regularly do a screening of the surrogacy clinics and appoint a committee to report on the deviations, if any.
- ✓ The surrogacy center should be registered with the State and Central governments whichever is applicable and thereby fake or unregistered centers can be checked.
- ✓ The minimum compensation should be fixed by the Government for surrogate mothers because it is a huge market and there is scope for exploiting the surrogate mothers.
- ✓ The Government can recognize the surrogate mothers as legal mothers of the child.
- ✓ During the pregnancy, the clinic or indented duos may facilitate the surrogate mothers, to meet the family members, on the grounds of human rights
- ✓ The child born through surrogacy, on reaching maturity may be entitled to access the identity of the surrogate mother, if he or she wishes to.
- ✓ The child born through surrogacy, must be presumed to be belonging to the commissioning couple. It may be made a mandate to provide post surrogacy care to the surrogate mother in need.
- ✓ The surrogate should get health insurance in case of health complications in post surrogacy, which must be made part of the agreement.
- ✓ The surrogate mothers and other stakeholders must not take surrogacy as commercial, which is not possible after the advent of New ART Bill, restricting commercial surrogacy. It must be treated as a service for which a nominal payment may be in order.
- ✓ The presence of intermediaries must be given up and those who wish to get a surrogate mother, may do so through advertisement or registering with a infertility treatment centre and get the child when their turn comes, like it is being followed for the adoption.
- ✓ The interests of both the intended and surrogate mothers must be safeguarded by fixing a fixed payment, uniformly applicable to all

clinics, without any deviation and the account must be maintained, showing to both the parties without any leakages. All these acts will help to maintain the service to become non-commercial.

Critique of the New ART Bill 2016 approved in March 2018 by Government of India

The Bill on Surrogacy seeks to prevent the abuses of the practice of surrogacy like unethical practices, disowning of children born out of surrogacy and mushrooming of intermediaries importing human embryos and gametes and exploitation of surrogates. The 228th report of the Law Commission of India has recommended, prohibiting commercial surrogacy, and allowing ethical altruistic surrogacy by enacting a suitable legislation.

- ✓ Accordingly, the Union Cabinet of Government of India approved the amended New ART (Regulation) Bill, 2016, during March 2018. The New Surrogacy bill was primarily concerned with preventing the exploitation of surrogate mothers, who are drawn from economically and socially backward background.
- ✓ It has highlighted the rights of a child, born through surrogacy and they must be treated equal to those of a natural or biological child. It emphasized the registration of surrogacy clinics. The new amended bill provided 16 months of extended insurance coverage, for surrogate mothers, to cover all complications, as well as a strict clause to safeguard the surrogate mother from exploitation. In fact the post surrogacy complications have not been taken seriously and no provision was there in the earlier bill.
- ✓ The Government of India reiterated that the new legislation on surrogacy, ensures effective regulation of surrogacy and prohibits commercial surrogacy and allows only altruistic surrogacy to infertile Indian couples. It is underlined that once it becomes an Act, it will regulate surrogacy services throughout the country, to control unethical practices in surrogacy and prevent the commercialization of surrogacy. It will also prohibit potential exploitation of surrogate mothers and children born through surrogacy. It says that "While commercial surrogacy will be prohibited, including sale and purchase of human embryo and gametes, ethical surrogacy to the needy infertile couples will be allowed on fulfillment of certain conditions and for specific purposes," All infertile Indian married couple, who want to avail ethical surrogacy, will be benefitted. Further, the rights of surrogate mother and children born out of surrogacy will be protected. So the focus of the new bill was on preventing the overseas couples to seek surrogacy and providing opportunities of altruistic surrogacy to Indian infertile couple by completely banning the commercial surrogacy.

- ✓ By excluding widows, live-in partners and homosexuals and imposing a "blanket ban" on foreigners, etc has been appearing a anti liberal and going back to stone age, without considering the emerging needs and scenario.
- ✓ Culture has come in the way of women's participation in public life and contribution to nation building. There is moral policing, by monitoring any violation of the customary norms related to marriages. The new forms and choice of life partnering is on the increase and live-in relationships and gender non-conformity groups are emerging, with LGBQT who want a child, through surrogacy process. Banning surrogacy to married person alone, will exclude gender non-conformity groups etc.
- ✓ It shuts the opportunity to those who wish to have their children despite their attempt to get child through adoption or through ARTs. On the other hand, it completely closes the opportunity of income earning through surrogacy. By banning instead of regulating, respecting the sentiments of the couples to have a child, it is equally losing foreign exchange earnings.
- ✓ The New Surrogacy Bill 2016 permits only married couples, to be eligible for surrogacy and excludes single parents, live-in couples and gays, etc. It is also underlined that the couple must certify that one of them is medically unfit to reproduce naturally. Couples, who have biological or adopted children, are not eligible for a surrogate. It was criticized by that "When the law allows adoption by a single parent, why not allow surrogacy too?." Indira Hinduja, a Mumbai-based infertility specialist, who delivered India's first test-tube baby in 1988, asked soon after the cabinet approved the bill.
- ✓ It is responded by the members who drafted the bill that "not that the government wants to completely ban the renting of wombs, but it wants to make it an altruistic practice, where eligible women offer to be surrogates for family members in need. It is heavily criticized by questioning that who and how many women will come forward to bear altruistic surrogacy, without any incentives or advantage for 10 months to bear a child for others. Will it not lead to harassment of women being forced by members of the family to serve as altruistic surrogate? Several doctors have also expressed their views on similar lines.
- ✓ The bill proposes a fine of Rs: 10 lakhs and jail terms of up to 10 years for violations. It is good that such stringent punishment is being imposed; But it must be monitored, which is often missing in the Indian Legal system.
- ✓ On the demand side, there would be revenue loss to the nation, by completely banning overseas surrogacy offers, which has been around 80% of the surrogate treatment in India. On the supply side, the poor women, who gave appointment and waiting for their turn to serve as

surrogate, had lost their earnings to feed their families. The Government is silent over such thousands of women who opted for surrogacy but continue to remain in debt, and could not feed their children. State cannot impose what the women should do on their bodies is the extreme form of criticism from the doctors and activists.

- ✓ The concerns of the activists are very much on the exploitation of the surrogate mothers and the bill must regulate the violation of their rights and not on safeguarding the cultural norms.
- ✓ The bill, applicable to the whole of India, except Jammu and Kashmir, proposes to regulate surrogacy in India, by establishing the National Surrogacy Board at the central level, State Surrogacy Boards and appropriate authorities in the states and union territories.
- ✓ The feminist critique is that it cannot just treat women's body as reproductive machine and at the same time, another school of thought says that the State cannot decide what the women should do with their bodies. It is their right to serve as surrogate for both economic and social reasons. It opens opportunities for several other categories of people, apart from infertile couple, to have a biological child. It has been bringing considerable revenue to the country.
- ✓ The new surrogacy bill 2018 must reconsider complete banning of commercial surrogacy and instead it could regulate and strictly monitor. Given that the new gender non-conforming groups are willing to get a child, they must be provided with the option to have one through surrogacy. Whether the State permits or not, live-in relationships are emerging and it turns out to be a constitutional right to choose their lives as per their wish. The proposed altruistic surrogacy will equally bring challenge and it would further lead to illegal surrogacy to continue where the surrogate mothers will still be exploited.

Suggestions on the New ART Bill

Based on the data gathered from multiple stakeholders the following suggestions, from gender and feminist perspectives, are placed with respect to regulation related to the surrogacy bill:

- ✓ The Government may need to initiate the 'Surrogacy Legal Center' like the one on blood donation and organ transplantation etc, to support the surrogate women and safeguard their rights and interests.
- ✓ Based on the information, the surrogate mothers and the needy couples must be given permission by the government, after assessing the need by a third party or a committee constituted to assess the genuine needs of such persons beyond any compromise.

- ✓ Needy people alone may be allowed for surrogacy process and after proper permission, the physician must start the surrogacy process.
- ✓ Transparency must be maintained and the report to the Government must be submitted once the contact is done.
- ✓ The payment details must be reported to the Government to inform that everything has been followed and fixed as per the law, including the remuneration in case the commercial surrogacy is admitted and regulated.
- ✓ Surrogate should be legal guardians of the child. The altruistic surrogacy can be entertained wherever the close relative is volunteering for and willing to.
- ✓ Commercial surrogacy must focus on human right and not exploitation and money making.
- ✓ The women's body is not just a reproductive machine but there is pain and labour behind every delivery, which can by no means be exploited in the name of paying money.

Even before the New Draft ART Bill 2016, the Center for Social Research (2012) suggested that there is need for concrete legal framework to monitor and regulate the surrogacy system and a clearly defined law needs to be drafted immediately which will pronounce in detail the Indian's government's stand on surrogacy so that exploitation of the surrogate mother can be stopped. Concurring with the CSR study, this study also strongly recommends, that instead of banning, commercial surrogacy should be regulated through a comprehensive compensation policy. An unregulated surrogacy industry, may function in violation of medical guidelines and not have enough legal provisions to safeguard the interests of all stakeholders and also legalizing surrogacy builds a win-win situation to all the stakeholders. It is also added that surrogate mother should be adequately and reasonably compensated and the compensation should be commensurate with all the costs she bears. This should include expenses related to post delivery care and child care support. Her kin should be compensated in case of her death during pregnancy or child birth. There is no legally approved payment structure for surrogate mothers and they are dependent on external members who facilitate surrogacy contracts between the stakeholders. So the government may consider the voice of the all, to legalize the commercial surrogacy. It is recommended in this study, to regulate with news provisions rather than completely banning surrogacy.

Reflections

There has been very limited research study on Surrogacy, as an ART, to produce a child for infertile couples. SAMA's study (2012) contextualizes surrogacy in broader debates around women's work and reproductive labour and legal regulatory framework on surrogacy. The ideological factors that drive the surrogacy industry which includes the stigma of infertility, interplay of 'natural' and 'biological' motherhood that legitimize surrogacy, using IVF in particular ways and the emerging market for reproductive technologies, have also been critically analyzed. It has asserted that there is a need for strengthening of linkages between academia and activism, that builds a perspective on the interaction of market, technology, patriarchy, and hetero-normativity, as seen in this practice (SAMA, 2012).

Council for Social Research (2012), in its study, provides a nuanced analysis of the motivations, experiences and consequences of surrogacy, that is rapidly proliferating in metro cities of India like Delhi and Mumbai. The study focuses on the reasons for the woman's decision to become a surrogate mother and for commissioning parents to opt for surrogacy. It provides a retrospective view of the relationship between commissioning parents and surrogate mother before the pregnancy, during the pregnancy, and after the birth. The study has provided the critical issues, that the surrogate mothers faced.

From the feminist Standpoint theory and gender socialization theory, it can be inferred that women have always been put into secondary position, which is true in the surrogacy context also. Surrogacy actually subscribes to the 'Centrality of Motherhood,' where the infertile couples are being forced to get a child and where the women bear the whole burden. Incidentally, agrees with in notion of "glorifying motherhood." Such perceptions have been reinforced in the socialization process, which the gender socialization theory has been emphasizing. The liberal feminists advocate that women can extend their labour, to economic participation in the market, with the equal focus on the reproductive role. Such views and perceptions, being the social norms, do not liberate women in the current context. In the surrogacy context, it is often raised that "When men donate sperm, why not the women donate egg and engage in surrogacy" because in a certain context, surrogacy has been viewed as equivalent to sex work. There is lack of scientific understanding on the process of surrogacy. Several women are kept in the dark about the treatment process, medicines, terms and conditions. Like any other technology, the ART and surrogacy is also a male generated technology and hence feminists point out that such ARTs reduced the female to a machine and baby to a product Hence it cannot be glorified that surrogacy or ART is a boon to the infertile couples as long as such technologies do reinforce women into the stereotype of motherhood and ill treat them if they do not reproduce. The present study has analyzed the data from the perspective of not reducing the women to mere reproduction and also on the needs of protecting the rights of surrogate mothers.

Conclusion

The present study on surrogacy and its connected issues, has brought out clearly various issues and challenges, involved with the surrogacy process and the kind of challenges and prospects that each stakeholder faced. There are various stakeholders involved on both the demand and supply side. On demand side, the intended parents are in desperate need for a child and hence willing to spend to get a child on their own and a biological child too. On the supply side, there are several stakeholders operate and they turn this surrogate service into a market. The feminist concerns are against treating the infertile woman as a curse and forcing them to get a child, by going through hardships, mental illness and incurring huge economic burden, which would have otherwise been available for the well being of the family. Adoption has not been encouraged much, which would partly take care of the pains of the infertile couples. The culture and customary norms, that woman must marry and if married must get a child and that too their own biological child, have engendered ARTs. From the feminist and gender perspective, the study thoroughly analyzed how surrogates had to convince and persuade their family members and also convince themselves to opt for surrogacy. The pain starts right from forcing the self to accept and it further aggravates when it is to be conveyed to the members of the family. In the process, the entire burden has been on the heads of women as if it is the duty and responsibility of women to arrange for the repayment of the debt or meet the livelihood needs of the households while male members spend their income on alcohol and add to accumulated debt. Every woman, in the sample, had to spend their time and energy to first offer to accept and later as surrogate mothers, particularly those who remained with intended parent's custody or relatives, had to face serious psychological issues. Post surrogacy care has been totally absent and there is no guarantee to surrogates for post surrogacy complications. The commercial surrogacy market did not have any regulation till 2014. The medical tourism on surrogacy has immensely helped the country to earn foreign exchange reserves but at the cost of several other issues including the individual women. There were other conflicts reported, particularly issues related to intended parents getting divorced during the surrogacy process. Hence it is good that the New ART Bill 2016 has been drafted and might become an Act soon. There are some issues with respect to the New ART bill 2014, particularly with respect to promotion of 'altruistic surrogacy' alone, where there will be other kinds of issues cropping up, and the relatives will be forced to opt for surrogacy though they may not be interested. Illegal commercial surrogacy would also be promoted, irrespective of the monitoring system, as it has been observed with sex selective abortions in spite of PCPNDT Act being implemented. But still legalizing surrogacy creates awareness in the society and saves surrogates from the exploitation, discrimination, oppression, suppression and marginalization of the women. The women underwent surrogacy, with all enthusiasm to experience or enjoy a better socio-economic and cultural life. But their expectations and dreams were largely unrealized due to the psychological, emotional, economic and physical exploitations by the various stakeholders, such as doctors, intermediates, family and society. To overcome all these issues, the Government, instead of banning, may legalize surrogacy with strong rules and regulations, with an inbuilt monitoring mechanism for the violations of the legal provisions. It is not to be entertained as a commercial and market based profession, where it only treats women's body as a reproductive machine which can never be encouraged. Women's interest, as human being, cannot be foregone for any requirements in the society and such feminist concerns must be respected forever.

Bibliography

- Abu-Rabia, A. (2013). Infertility and Surrogacy in Islamic Society: Socio-Cultural, Psychological, Ethical, and Religious Dilemmas. *Open Psychology Journal*, *6*, 54-60.
- Ainsworth, S. L. (2015). Bearing children, bearing risks: feminist leadership for progressive regulation of compensated surrogacy in the United States.
- Akin, M. (2005). Assisted Reproductive Technology and the Law.
- Alaro, A. A., Berna, A., & Vardit, R. C. (2012). Assisted reproductive technology (ART): the Islamic Law perspective. *Islam and bioethics. Ankara University*, *Turkey*, 95-108.
- Allen, A. L. (1990). Surrogacy, slavery, and the ownership of life. *Harv. JL & Pub. Pol'y*, *13*, 139.
- Allen, A. L. (1991). The black surrogate mother. *Harv. BlackLetter J.*, *8*, 17.
- Almond, B. (2008). Reproductive Choice: Whose Rights? Whose Freedom?. *The Contingent Nature of Life*, 169-182.
- Alsgaard, H. (2012). Decoupling Marriage & Procreation: A Feminist Argument for Same-Sex Marriage. *Berkeley J. Gender L. &Just.*, *27*, 307.
- Andersen, A. N., Goossens, V., Gianaroli, L., Felberbaum, R., De Mouzon, J., &Nygren, K. G. (2007). Assisted reproductive technology in Europe, 2003. Results generated from European registers by ESHRE. *Human Reproduction*, *22*(6), 1513-1525.
- Anderson, E. S. (2000). Why commercial surrogate motherhood unethically commodifies women and children: reply to McLachlan and Swales. *Health care analysis*, *8*(1), 19-26.
- Andrews, L. B. (1995). Beyond doctrinal boundaries: a legal framework for surrogate motherhood. *Virginia Law Review*, 2343-2375.
- Armour, K. L. (2012). An overview of surrogacy around the world: trends, questions and ethical issues. *Nursing for women's health*, *16*(3), 231-236.
- Atwell, B. L. (1989). Surrogacy and adoption: a case of incompatibility.

- Banerjee, A. (2011). *Re-conceiving "borders": A feminist pragmatic phenomenology for postcolonial feminist ethics and politics* (Doctoral dissertation, University of Oregon).
- Banerjee, S. (2012). Emergence of the 'surrogacy industry.' *Economic and political weekly*, *47*(11), 27-29.
- Banerjee, S. (2013). Gestational surrogacy contracts: Altruistic or commercial? A contract theoretic approach. *The Manchester School*, *81*(3), 438-460.
- Banerjee, S., &Basu, S. (2009). Rent a womb: surrogate selection, investment incentives and contracting. *Journal of Economic Behavior & Organization*, *69*(3), 260-273.
- Baumhofer, E. (2012). Commodifying the Female Body: Outsourcing Surrogacy in a Global Market.
- Bee, A. R. Assisted Reproductive Technologies Abuse Women in the form of Surrogacy. International Journal of Management Research and Social Science (IJMRSS) ISSN 2394-6407(Print) Volume 2, Issue 4, October – December 2015 ISSN 2394-6415(Online)
- Bee, A.R. (2016). Surrogate Mother - Rights and Challenges (I) 09-15.
- Bender, L. (1992). Overview of Feminist Torts Scholarship. *Cornell L. Rev.*, *78*, 575.
- Ber, R. (2000). Ethical issues in gestational surrogacy. *Theoretical Medicine and Bioethics*, *21*(2), 153-169.
- Berra, M., Liao, L. M., Creighton, S. M., & Conway, G. S. (2010). Long-term health issues of women with XY karyotype. *Maturitas*, *65*(2), 172-178.
- Bewkes, F. J. (2014). Surrogate or Mother: The Problem of Traditional Surrogacy. *Tenn. J. Race Gender & Soc. Just.*, *3*, 143.
- Bhalla, N., &Thapliyal, M. (2013). India seeks to regulate its booming 'rent-a-womb' industry. *Mon Sep*, *30*, 1.
- Bharadwaj, A. (2003). Why adoption is not an option in India: the visibility of infertility, the secrecy of donor insemination, and other cultural complexities. *Social science & medicine*, *56*(9), 1867-1880.
- Blatt, R. (2009). Wombs for Rent: Gestational Surrogacy and the New Intimacies of the Global Market. *Bachelor of Arts Hons. Paper, Brown University.*
- Bowdin, S., Allen, C., Kirby, G., Brueton, L., Afnan, M., Barratt, C., ..& Reardon, W. (2007). A survey of assisted reproductive technology births and imprinting disorders. *Human Reproduction*, *22*(12), 3237-3240.
- Boyd, E. A., Lo, B., Evans, L. R., Malvar, G., Apatira, L., Luce, J. M., & White, D. B. (2010). "It's not just what the doctor tells me": Factors

that influence surrogate decision-makers' perceptions of prognosis. *Critical care medicine*, *38*(5), 1270.

- Brandel, A. (1995). Legislating surrogacy: a partial answer to feminist criticism. *Md. L. Rev.*, *54*, 488.
- Campbell, A. V. (2000). Surrogacy, rights and duties: a partial commentary. *Health care analysis*, *8*(1), 35-40.
- Council for Social Research (2012) Research-Studies-Surrogacy-Motherhood-Ethical-or-Commercial-Delhi&Mumbai. Pdf http://www.csrindia.org/surrogate-motherhood).
- Chamberlain, K. (2000). Methodolatry and qualitative health research. *Journal of Health Psychology*, *5*(3), 285-296.
- Chatterjee, M. P. (2016). Introducing a New Form of Bonded Labour Arising from Commercial Surrogacy:-Special Reference to India. *Paripex- Indian Journal of Research*, *4*(12).
- Chaturvedi, R. K., Garg, S. P., Mishra, A., Garg, V., &Chaturvedi, P. (2014). Surrogacy policy in India and need of acts to regulate commercial surrogacy. *Journal of evolution of medical and dental sciences*, *3*(20), 5384-5389.
- Cherry, A. L. (2014). The Rise of the Reproductive Brothel in the Global Economy: Some Thoughts on Reproductive Tourism, Autonomy, and Justice. *U. Pa. JL & Soc. Change*, *17*, 257.
- Ciccarelli, J. C., & Beckman, L. J. (2005). Navigating rough waters: an overview of psychological aspects of surrogacy. *Journal of Social Issues*, *61*(1), 21-43.
- Cook, R., Sclater, S. D., &Kaganas, F. (2003). *Surrogate motherhood: international perspectives*. Hart Publishing.
- Damelio, J., & Sorensen, K. (2008). Enhancing autonomy in paid surrogacy. *Bioethics*, *22*(5), 269-277.
- Dar, S., Lazer, T., Swanson, S., Silverman, J., Wasser, C., Moskovtsev, S. I., ...&Librach, C. L. (2014). Assisted reproduction involving gestational surrogacy: an analysis of the medical, psychosocial and legal issues: experience from a large surrogacy program. *Human reproduction*, *30*(2), 345-352.
- Dashiell, C. (2012). From Louise Brown to Baby M and Beyond: A Proposed Framework for Understanding Surrogacy. *Rutgers L. Rev.*, *65*, 851.
- Davies, M. J., Moore, V. M., Willson, K. J., Van Essen, P., Priest, K., Scott, H., & Chan, A. (2012). Reproductive technologies and the risk of birth defects. *New England Journal of Medicine*, *366*(19), 1803-1813.
- De Brucker, M., Tournaye, H., Haentjens, P., Verheyen, G., Collins, J., & Camus, M. (2013). Assisted reproduction counseling in women aged

40 and above: a cohort study. *Journal of assisted reproduction and genetics*, *30*(11), 1431-1438.

- DeMarco, J. P., Powell, D. P., & Stewart, D. O. (2011). Best interest of the child: Surrogate decision making and the economics of externalities. *Journal of Bioethical Inquiry*, *8*(3), 289-298.
- Deonandan, R., Green, S., & Van Beinum, A. (2012). Ethical concerns for maternal surrogacy and reproductive tourism. *Journal of Medical Ethics*, medethics-2012.
- Dickenson, D. L. (2001). Property and women's alienation from their own reproductive labour. *Bioethics*, *15*(3), 205-217.
- Edelmann, R. J. (2004). Surrogacy: the psychological issues. *Journal of Reproductive and Infant Psychology*, *22*(2), 123-136.
- Ergas, Y. (2012). The Transnationalization of Everyday Life: Cross-border Reproductive Surrogacy, Human Rights and the Re-visioning of International Law. *Columbia University*.
- Ezra, o. (2006). Procreation after death. *Moral dilemmas in real life*, 115-126.
- Fadel, H. E. (2002). The Islamic viewpoint on new assisted reproductive technologies. *Fordham Urb. LJ*, *30*, 147.
- Fasouliotis, S. J., &Schenker, J. G. (1999). Social aspects in assisted reproduction. *Human Reproduction Update*, *5*(1), 26-39.
- Gaard, G. (2010). Reproductive technology, or reproductive justice?: An eco feminist, environmental justice perspective on the rhetoric of choice. *Ethics & the Environment*, *15*(2), 103-129.
- Gabry, L. I. (2011). Procreating without pregnancy: Surrogacy and the need for a comprehensive regulatory scheme. *Colum. JL & Soc. Probs.*, *45*, 415.
- Galbraith, M., McLachlan, H. V., & Swales, J. K. (2005). Commercial agencies and surrogate motherhood: a transaction cost approach. *Health Care Analysis*, *13*(1), 11-31.
- Galpern, E. (2007). Assisted reproductive technologies: Overview and perspective using a reproductive justice framework. *Center for Genetics and Society*, *Oakland, CA*, *20*.
- Ghosh, A. (2015). Amatory fiction of Eliza Haywood. Strathern, M. (2003). Still giving nature a helping hand? Surrogacy: A debate about technology and society. *Surrogate motherhood: International perspectives*, 281-296.
- Gigerenzer, G. (1998). Surrogates for theories. *Theory & Psychology*, *8*(2), 195-204.
- Goud, M. S., &Sunkara, A. (2012). Is Legalizing Surrogacy-An Outsourcing Motherhood?. *International Journal of Advancements in Research & Technology*, *1*(4), 98-106.

- Gugucheva, M. (2010). Surrogacy in America. *Cambridge, MA: Council for Responsible Genetics Cambridge*, 1-40.
- Gupta, J. A., &Richters, A. (2008). Embodied subjects and fragmented objects: Women's bodies, assisted reproduction technologies and the right to self-determination. *Journal of Bioethical Inquiry*, *5*(4), 239-249.
- Hacker, D. (2001). Single and married women in the law of Israel–A feminist perspective. *Feminist Legal Studies*, *9*(1), 29-56.
- Hansen, M., Bower, C., Milne, E., de Klerk, N., & J. Kurinczuk, J. (2005). Assisted reproductive technologies and the risk of birth defects—a systematic review. *Human reproduction*, *20*(2), 328-338.
- Hendricks, J. S. (2006). Essentially a Mother. *Wm. & Mary J. Women & L.*, *13*, 429.
- Hibino, Y., &Shimazono, Y. (2013). Becoming a Surrogate Online: "Message Board" Surrogacy in Thailand. *Asian Bioethics Review*, *5*(1), 56-72.
- Hofman, D. L. (2008). Mama's Baby, Daddy's Maybe: A State-by-State Survey of Surrogacy Laws and Their Disparate Gender Impact. *Wm. Mitchell L. Rev.*, *35*, 449.
- Ingram, J. D. (1992). Surrogate Gestator: A New and Honorable Profession. *Marq. L. Rev.*, *76*, 675.
- Jadva, V., &Imrie, S. (2013). Children of surrogate mothers: psychological well-being, family relationships and experiences of surrogacy. *Human Reproduction*, *29*(1), 90-96.
- Jadva, V., Imrie, S., &Golombok, S. (2014). Surrogate mothers 10 years on: a longitudinal study of psychological well-being and relationships with the parents and child. *Human reproduction*, *30*(2), 373-379.
- Kaczor, C. (2007). Artificial Wombs and Embryo Adoption. In *The Ethics of Embryo Adoption and the Catholic Tradition* (pp. 307-322). Springer Netherlands.
- Kamel, R. M. (2013). Assisted reproductive technology after the birth of Louise brown. *Journal of reproduction & infertility*, *14*(3), 96.
- Karkal, M. (1997). Surrogacy from a feminist perspective. *Indian Journal of medical Ethics*, *Oct-Dec*, *5*(4).
- Katz, P., Nachtigall, R., &Showstack, J. (2002). The economic impact of the assisted reproductive technologies. *Nature Medicine*, *8*(10; SUPP), 29-32.
- Kelly, B., Rid, A., &Wendler, D. (2012). Systematic review: individuals' goals for surrogate decision-making. *Journal of the American Geriatrics Society*, *60*(5), 884-895.

- Koob, M. (2015). Book review: wombs in labor: transnational commercial surrogacy in India by Amrita Pande. *LSE Review of Books*.
- Kumar, P., Inder, D., & Sharma, N. (2013). Surrogacy and women's right to health in India: Issues and perspective. *Indian journal of public health*, *57*(2), 65.
- Laufer-Ukeles, P. (2013). Mothering for money: Regulating commercial intimacy.
- Leibowitz-Dori, I. (1997). Womb for rent: The future of international trade in surrogacy. *Minn. J. Global Trade*, *6*, 329.
- Lieber, K. B. (1992). Selling the Womb: Can the Feminist Critique of Surrogacy Be Answered. *Ind. LJ*, *68*, 205.
- Lim, D. A., Chan, M. F., & Childs, C. (2013). Surrogate consent for critical care research: exploratory study on public perception and influences on recruitment. *Critical Care*, *17*(1), R5.
- Lim, H. (1999). Caesareans and cyborgs. *Feminist Legal Studies*, *7*(2), 133-173.
- Lin, T. (2012). Born Lost: Stateless Children in International Surrogacy Arrangements. *Cardozo J. Int'l & Comp. L.*, *21*, 545.
- Lori B. Andrews, & American Bar Foundation. (1987). *Feminist perspectives on reproductive technologies* (No. 8701). American Bar Foundation.
- Luk, J., Greenfeld, D. A., &Seli, E. (2010). Third party reproduction and the aging couple. *Maturitas*, *66*(4), 389-396.
- Macdonald, C. L. (1998). Manufacturing motherhood: The shadow work of nannies and au pairs. *Qualitative Sociology*, *21*(1), 25-53.
- Machinton, N. W. (1987). Surrogate Motherhood: Boon or Baby-Selling the Unresolved Questions. *Marq. L. Rev.*, *71*, 115.
- Majumdar, A. (2013). Transnational surrogacy: The "public" selection of selective discourse. *Economic and Political Weekly*, *48*(45-46), 24-27.
- Malik, R. (2015). Commercial surrogates in India: victims of globalization?. *South Asia@ LSE*.
- Martin, E. (1991). The egg and the sperm: How science has constructed a romance based on stereotypical male-female roles. *Signs: Journal of Women in Culture and Society*, *16*(3), 485-501.
- Marwah, V. (2011). Reinventing reproduction, re-conceiving challenges: an examination of assisted reproductive technologies in India. *Economic and political weekly*, 104-111.
- Mathur, K. (2008). Body as space, body as site: Bodily integrity and women's empowerment in India. *Economic and Political Weekly*, 54-63.

- McLachlan, H. V. (2002). Persons and their bodies: how we should think about human embryos. *Health Care Analysis*, *10*(2), 155-164.
- McLachlan, H. V., & Swales, J. K. (2000). Babies, child bearers and commodification: Anderson, Brazier et al., and the political economy of commercial surrogate motherhood. *Health Care Analysis*, *8*(1), 1-18.
- McLachlan, H. V., & Swales, J. K. (2001). Surrogate motherhood, rights and duties: A reply to Campbell. *Health Care Analysis*, *9*(1), 101-107.
- McLachlan, H., & Swales, J. (2005). Surrogate Motherhood: Beyond the Warnock and the Brazier Reports. *Human Reproduction & Genetic Ethics*, *11*(1), 12-23.
- Millbank, J. Unlikely Fissures and Uneasy Resonances: Lesbian Co-Mothers, Surrogate Parenthood and Fathers' Rights' (2008). *Feminist Legal Studies*, *16*, 141.
- Moorthy, A. (2011). *Transnational Commercial Gestational Surrogacy: Cultural Constructions of Motherhood and their Role in the Development of National Indian Guidelines* (Doctoral dissertation, Arizona State University).
- Mortazavi, S. (2011). It takes a village to make a child: creating guidelines for international surrogacy. *Geo. LJ*, *100*, 2249.
- Munro, V. E. (2001). Surrogacy and the construction of the maternal-foetal relationship: the feminist dilemma examined. *Res Publica*, *7*(1), 13-37.
- Nelson, H. L., & Nelson, J. L. (1992). Cutting Motherhood in Two: Some Suspicions Concerning Surrogacy.' *Feminist Perspectives in Medical Ethics*, (695), 257.
- Neyer, G., &Bernardi, L. (2011). Feminist perspectives on motherhood and reproduction. *Historical Social Research/HistorischeSozial forschung*, 162-176.
- Norton, W., Hudson, N., &Culley, L. (2013). Gay men seeking surrogacy to achieve parenthood. *Reproductive biomedicine online*, *27*(3), 271-279.
- Nussbaum, M. C. (1998). "Whether from reason or prejudice": taking money for bodily services. *The Journal of Legal Studies*, *27*(S2), 693-723.
- Pande, A. (2009). Not an 'Angel,' not a 'Whore' Surrogates as 'Dirty' Workers in India. *Indian Journal of Gender Studies*, *16*(2), 141-173.
- Pande, A. (2010). "At Least I Am Not Sleeping with Anyone": Resisting the Stigma of Commercial Surrogacy in India. *Feminist Studies*, *36*(2), 292-312.

- Pande, A. (2010). Commercial surrogacy in India: Manufacturing a perfect mother-worker. *Signs: Journal of Women in Culture and Society*, *35*(4), 969-992.
- Pande, A. (2011). Transnational commercial surrogacy in India: gifts for global sisters?. *Reproductive biomedicine online*, *23*(5), 618-625.
- Pande, A. (2016). Global reproductive inequalities, neo-eugenics and commercial surrogacy in India. *Current Sociology*, *64*(2), 244-258.
- Panitch, V. (2013). Global surrogacy: Exploitation to empowerment. *Journal of Global Ethics*, *9*(3), 329-343.
- Panitch, V. (2013). Surrogate tourism and reproductive rights. *Hypatia*, *28*(2), 274-289.
- Parkinson, J., Tran, C., Tan, T., Nelson, J., Batzofin, J., &Serafini, P. (1999). Perinatal outcome after in-vitro fertilization-surrogacy. *Human Reproduction*, *14*(3), 671-676.
- Paxson, H. (2006). Reproduction as spiritual kin work: orthodoxy, IVF, and the moral economy of motherhood in Greece. *Culture, medicine and psychiatry*, *30*(4), 481-505.
- Pelka, S. (2009). Third-Party Reproduction: Creating Kinship through an Intent to Parent. *Anthropology News*, *50*(2), 8-12.
- Pelzman, J. (2010). "Womb for Rent" Gestational Surrogacy Contracts-A New Path for Outsourcing Service Contracts.
- Peng, L. (2012). Surrogate mothers: an exploration of the empirical and the normative. *Am. UJ Gender Soc. Pol'y& L.*, *21*, 555.
- Points, K. (2009). Commercial surrogacy and fertility tourism in India. *Durham, NC: Kenan Institute of Ethics, Duke University. Retrieved August*, *1*, 2011.
- Polkinghorne, D. E. (1988). Narrative knowing and the human sciences. Albany, NY: State University of New York Press.
- Polkinghorne, D. E. (2000). Psychological inquiry and the pragmatic and hermeneutic traditions. *Theory and Psychology*, *10*(4), 453-479.
- Poote, A. E., & van den Akker, O. B. (2009). British women's attitudes to surrogacy. *Human reproduction*, *24*(1), 139-145.
- Prasad, M., &Shekhawat, S. (2014). Arbitrability of Cross-Border Surrogacy Disputes: An Indian Perspective. *Chartered Accountant*, *22*.
- Qadeer, I., & John, M. E. (2009). The business and ethics of surrogacy. *Economic and political weekly*, 10-12.
- Raymond, J. G. (1989). At Issue Reproductive Technologies, Radical Feminism, and Socialist Liberalism.
- Readings, J., Blake, L., Casey, P., Jadva, V., &Golombok, S. (2011). Erratum to "Secrecy, disclosure and everything in-between: decisions of parents of children conceived by donor insemination, egg donation

and surrogacy" [Reprod. BioMed. Online 22 (2011) 485–495]. *Reproductive Biomedicine Online*, *22*(6), 653-654.

- Recht, S. M. (1987). M is or Money: Baby M and the Surrogate Motherhood Controversy. *Am. UL Rev.*, *37*, 1013.
- Reissman, C.K. (1993). Narrative analysis. *Qualitative Research Methods*, *30*, 1-17.
- Rhodes, R., &Holzman, I. (2004). The not unreasonable standard for assessment of surrogates and surrogate decisions. *Theoretical medicine and bioethics*, *25*(4), 367-386.
- Ricciarelli, E., Bruna, I., Verdú, V., Torrelló, M. J., Herrer, R., Gris, J. M., ...& Cabello, Y. (2013). Impact of assisted reproduction treatments on Spanish newborns: report of 14,119 pregnancies. *Journal of assisted reproduction and genetics*, *30*(7), 897-905.
- Roosevelt III, K. (1998). The Newest Property: Reproductive Technologies and the Concept of Parenthood. *Santa Clara L. Rev.*, *39*, 79.
- Rothman, B. K. (1991). Reproductive technologies and surrogacy: A feminist perspective. *Creighton L. Rev.*, *25*, 1599.
- Roye, C. F., & Balk, S. J. (1996). The relationship of partner support to outcomes for teenage mothers and their children: A review. *Journal of Adolescent Health*, *19*(2), 86-93.
- Ryan, M., & Berkowitz, D. (2009). Constructing gay and lesbian parent families "beyond the closet." *Qualitative sociology*, *32*(2), 153-172.
- SAMA- Research group for Women, S. R. G. (2008). Assisted reproductive technologies: Autonomy or subjugation? A case study from India. In *Women's Studies International Forum* (Vol. 31, No. 4, pp. 319-325).
- SAMA (2012) "Birthing A Market: A Study on Commercial Surrogacy," Research Publication.
- Samuelson, P. A. (1962). Parable and realism in capital theory: the surrogate production function. *The Review of Economic Studies*, *29*(3), 193-206.
- Saravanan, S. (2013). An ethno methodological approach to examine exploitation in the context of capacity, trust and experience of commercial surrogacy in India. *Philosophy, Ethics, and Humanities in Medicine*, *8*(1), 10.
- Sarojini, N., Marwah, V., &Shenoi, A. (2011). Globalization of birth markets: a case study of assisted reproductive technologies in India. *Globalization and health*, *7*(1), 27.
- Satz, D. (2010). Ethics, economics, and markets: An interview with Debra Satz. *Erasmus Journal for Philosophy and Economics*, *3*(1), 68-88.

- Schanbacher, K. (2014). India's Gestational Surrogacy Market: An Exploitation of Poor, Uneducated Women. *Hastings Women's LJ*, *25*, 201.
- Scott, E. S. (2009). Surrogacy and the Politics of Commodification. *Law and Contemporary Problems*, *72*(3), 109-146.
- Sera, J. M. (1996). Surrogacy and prostitution: A comparative analysis. *Am. UJ Gender & L.*, *5*, 315.
- Shalev, C. (2012). An ethic of care and responsibility: reflections on third-party reproduction. *Medicine Studies*, *3*(3), 147-156.
- Shanley, M. L. (1993). "Surrogate Mothering" and women's freedom: A critique of contracts for human reproduction. *Signs: Journal of Women in Culture and Society*, *18*(3), 618-639.
- Sharma, R. (2007). An International, Moral & Legal Perspective: The Call for Legalization of Surrogacy in India.
- Sky, I. (2011). *Motherhoods, Bodies and Inequalities: An Exploration of Surrogacy and Its Implications For Social Work* (Doctoral Dissertation).
- Smerdon, U. R. (2008). Crossing bodies, crossing borders: International surrogacy between the United States and India. *Cumb. L. Rev.*, *39*, 15.
- Spar, D. (2005). For love and money: the political economy of commercial surrogacy. *Review of international political economy*, *12*(2), 287-309.
- Sri, B. S. (2010). Women's Bodies and the Medical Profession. *Economic and Political Weekly*, 52-57.
- Stanmeyer, S. (2007). An Embryo Adoptive Mother's Perspective. In *The Ethics of Embryo Adoption and the Catholic Tradition* (pp. 237-249). Springer Netherlands.
- Storrow, R. F. (2013). New thinking on commercial surrogacy. *Ind. LJ*, *88*, 1281.
- Strathern, M. (1995). Future kinship and the study of culture. *Futures*, *27*(4), 423-435.
- Strathern, M. (2003). Still giving nature a helping hand? Surrogacy: A debate about technology and society. *Surrogate motherhood: International perspectives*, 281-296.
- Stteinbock, B. (1996). Regulating assisted reproductive technologies: An ethical framework. *Women's Health Issues*, *6*(3), 167-174.
- Svitnev, K. (2011). 13 Legal control of surrogacy–international perspectives.
- Tashi, S., Mehran, N., Eskandari, N., & Tehrani, T. D. (2014). Emotional experiences in surrogate mothers: A qualitative study. *Iranian journal of reproductive medicine*, *12*(7), 471.

- Teman, E. (2008). The social construction of surrogacy research: An anthropological critique of the psychosocial scholarship on surrogate motherhood. *Social science & medicine*, *67*(7), 1104-1112.
- Teman, E. (2010). *Birthing a mother: The surrogate body and the pregnant self.* Univ of California Press.
- Thapa, J. D. (2010). The Babies M: The Relevance of Baby Manji Yamada v. Union of India (UOI) and in the Matter of Baby M. *J. Indian L. &Soc'y*, *2*, 83.
- Thapa, J. D. (2012). *Analyzing the Status of the Surrogate Mother under the Assisted Reproductive Technologies (Regulation) Bill, 2010.* NUJS Working Paper Series, the West Bengal National University of Juridical Sciences.
- Tieu, M. (2007). Oh Baby: The Problem of Surrogacy. *Bioethics Research Notes*, *19*(1), 1-9.
- Tinsman, H. (1992). Behind the Sexual Division of Labor: Connecting Sex to Capitalist Production. *Yale J. Int'l L.*, *17*, 241.
- Trowse, P. (2011). The Commercial Surrogacy Dilemma. *The Queensland Lawyer*, *31*(3), 160-162.
- Truong, T. D. (2015). Human trafficking, globalization, and transnational feminist responses. *The Oxford Handbook of Transnational Feminist Movements*, 295.
- Unnithan, M. (2013). Thinking through surrogacy legislation in India: reflections on relational consent and the rights of infertile women. *Journal of Legal Anthropology*, *1*(3), 287-313.
- Van den Akker, O. (2000). The importance of a genetic link in mothers commissioning a surrogate baby in the UK. *Human reproduction*, *15*(8), 1849-1855.
- Van den Akker, O. B. (2007). Psychosocial aspects of surrogate motherhood. *Human reproduction update*, *13*(1), 53-62.
- Van Niekerk, A., & Van Zyl, L. (1995). The ethics of surrogacy: women's reproductive labour. *Journal of medical ethics*, *21*(6), 345-349.
- Van Zyl, L. (2002). Intentional parenthood and the nuclear family. *Journal of Medical Humanities*, *23*(2), 107-118.
- Van Zyl, L. (2008). A case for paid surrogate motherhood. In *Bearing and Rearing Children: The Ethics of Procreation and Parenthood.*
- Van Zyl, L., & Van Niekerk, A. (2000). Interpretations, perspectives and intentions in surrogate motherhood. *Journal of Medical Ethics*, *26*(5), 404-409.
- Voskoboynik, K. (2016). Clipping the Stork's Wings: Commercial Surrogacy Regulation and its Impact on Fertility Tourism. *Ind. Int'l & Comp. L. Rev.*, *26*, 336.

- Walks, M. (2007). Breaking the Silence.
- Wallbank, J. (2002). Too many mothers? Surrogacy, kinship and the welfare of the child. *Medical law review*, *10*(3), 271-294.
- Waugh, E. H. (1999). The Islamic Tradition: Religious Beliefs and Health Care Decisions.
- Weisberg, D. K. (2005). The birth of surrogacy in Israel.
- Weiss, R. S. (1994). *Learning from strangers. The art and method of qualitative* interview *studies*. New York, NY: The Free Press.\
- Whittaker, A. (2011). Cross-border assisted reproduction care in Asia: implications for access, equity and regulations. *Reproductive health matters*, *19*(37), 107-116.
- Whitton, L., &Frolik, L. A. (2011). Surrogate Decision-Making Standards for Guardians: Theory and Reality.
- Wilder, B. L. (2002). Assisted reproduction technology: trends and suggestions for the developing law. *J. Am. Acad. Matrimonial Law.*, *18*, 177.
- Wilkinson, S. (2003). The exploitation argument against commercial surrogacy. *Bioethics*, *17*(2), 169-187.
- York, S. G. (1990). A Contractual Analysis of Surrogate Motherhood and a Proposed Solution. *Loy. LAL Rev.*, *24*, 395.
- Zandi, M., Vanaki, Z., Shiva, M., Mohammadi, E., Bagheri-Lankarani, N., &Karimi, M. (2014). Legal constraints of using surrogacy. *Journal of Nursing and Midwifery Sciences*, *1*(3), 11-23.
- Zil, C. E. C. P., & Crawford, V. (2006). The Effects of Compensation on the Supply of Surrogate Mothers.

Web Sources

- http://theconversation.com/reject-commercial-surrogacy-as-another-form-of-human-trafficking-30314
- http://www.babycenter.com/0_fertility-treatment-assisted-reproductive-technologies-art_4093.bc
- http://www.cbsnews.com/news/possible-fertility-drug-cancer-link-found/ id
- http://www.dinakaran.com/News_Detail.asp?Nid=137966
- http://www.maalaimalar.com/2015/03/06142134/mother-changed-for-rental-moth.html
- http://www.tamilmurasu.org/inner_tamil_news.asp?Nid=58399
- http://www.bbc.com/news/uk-england-34467613
- http://ibnlive.in.com/news/is-surrogacy-leading-to-exploitation-of-poor-women/266801-3-118.html

- www.tamil.webdunia.com/printerfriendlyversion?article_id=11510280 0044_1
- https://samawomenshealth.wordpress.com/2013/08/
- http://www.ctvnews.ca/health/surrogates-feel-hurt-by-india-s-ban-on-foreign-customers-1.2663609
- http://www.thehindu.com/sci-tech/health/policy-and-issues/beautiful-and-fair-preferred-amongsurrogate-mothers-too/article4028640.ece
- http://v4news.com/surrogate-mother-delivered-grandchild-science-magic-happened-in-tamilnadu/
- http://www.newindianexpress.com/cities/chennai/Grandmother-Gives-Birth-to-her-Own-Granddaughter/2015/03/07/article2701384.ece
- https://www.google.co.in/?gfe_rd=cr&ei=Ud0nVf62NzA8gePsYCAB w#q=grandma+deliver+grandchild+by+surrogacy
- http://www.dailymail.co.uk/news/article-2963277/British-mum-59-bid-world-medical-ll-birth-baby-dead-daughter.html
- http://www.dinamalar.com/news_detail.asp?id=1174402
- http://www.thehindu.com / todays-paper / tp-national / tp-tamilnadu/ chennaibased-trust-seeks-to-bring-about-transparency-in-surrogacy /article4325195.ece
- http:// epaper.dinamani.com / 133046 / Dinamnai-Tiruchy/07-07-2013# page/8/1
- http://www.deccanchronicle.com / 130313 /news-current-affairs/article/ surrogate-moms-need-legal-help
- http://www.thehindu.com /sci-tech / health/policy-and-issues/beautiful-and-fair-preferred-among-surrogate-mothers-too/article4028640.ece ("beautiful and fair mother")
- http://talkmag.in/cms/trends/item/204-mommy-tourism
- http://articles.economictimes.indiatimes.com/2015-11-08/news/681089 25_1_hari-g-ramasubramanian-surrogacy-reproductive-technology-bill

Appendix I

Case Studies

Case: 1

Ms. Jeyanthi (Name Changed), Age: 28. She was from Viluppuram, four years before she lost her husband due to alcoholism. They are four in the family such as herself, two younger sisters and her mom. She has male and female child. She was working as a cook, at that time she came to know surrogacy through her mom's sister.

Her aunt son was student of the police training institute and often they are conducting parents meeting. Her aunt often went to the training institute to attend parents meeting, in that meeting they asked to refer someone to practice surrogacy for them; that is first time her aunt came to know about surrogacy. Because, the director of the institute lost his single son from accident with such background her aunt shared their thoughts with her and she asked her to help them. She doesn't have male members in her family to earn and taking care the family and also huge poverty; she decided to practice surrogacy with her aunt support. Except her aunt, none of her family not known that she underwent surrogacy. After discussion with the intended parents, they took her into hospital at Chennai. In the hospital they took complete check up and at last they declare that she was fit to do surrogacy. Before her, they had tried more women's but she only medically fit to start process.

To stay they are arranged a rented home in Chennai till the delivery. As per the process, they started to check her blood, urine, uterus etc. They are not collected the egg from her and also they got the egg donation from some other women because due to age bar of the intended mother. They got the signature from her and her aunt in the agreement. The contract was both in English and vernacular language but she is not able understand the clauses mentioned in the contract at the same time she read that, during surrogacy process if anything happen she is the solely responsible for everything.

She mentioned that, the intended parents shared to her, in this process the amount was fixed by the doctors alone. There is no intermediate between them and she is not demanded the amount. Before started the process they fixed the amount of Rs: 2.5L and the fixed amount was not yet settled by the intended parents after delivered they baby. She has delivered the baby from three months during the time of interview (June, 2015). Also, they gave the partial amount as few installments such as first installment of Rs: 35,000, then they will send monthly Rs: 3,000 to her family, incase if her children health is not good at the time they will send Rs: 1000 extra for hospital expenditure. Totally yet she has received 1.5L remaining balance not yet settled. She continuously asking them to settle the amount but the intended parents refuse to settle the money because they are mentioned that, they don't have money will give later. Even after a few months also they are not ready to settle the amount they are still asking her to wait and bargaining with her.

She can't able to do normal day today routine works which she has done before surrogacy. She mentioned that, always hey body was paining and feel tired and before surrogacy she was active but after surrogacy her health was completely affected. Now her family again facing economic problem so she plans to work but her health not supported to work.

In Chennai, Avavdi they arranged a rented home for staying, in that home she stayed alone up to five months. During this period she takes care herself and she prepared food for her which she likes to eat. Initially they are arranged a servant to serve her but she refused the servant because the servant may question her and she may be reveals the truth in future. During the process they asked her to take the complete rest but she cannot simply sit so she has done all her domestic chores till five months. After five months they intended parents arranged relatives to take care of her, to prepare the food and serve her. But she doesn't like the food which was prepared by them; due to this she lost her weight. That food was prepared by them and forced to eat her even she has nausea. Every time they are forced her to eat more and to do all the domestic works. Initially they treated her well later they are restricted to speak with the family members over phone and they are tortured her and abused her verbally because they thought she was going to deliver the baby for money. She had not undergone surrogacy for money, her prime motive was to help the childless couple and fulfill their desire. During the pregnancy her sister partner passed away for the condolence they are not allowed her to go and attend and also her son got injured in school and his ear was ripped for that also they are not allowed her to go her home. The intended parents allowed to go home once, during three months of pregnancy after that they are not allowed to go home. In this scenario she doesn't know what to do and she cannot share her feelings to the family members because of the confidentiality so she just convinced herself till the delivery because she cannot break the contract.

For IVF, they said the physician put two embryos (ET) and one embryo only fixed in the uterus because of her activities one embryo has spoiled and the intended parents blamed her and harmed her. To settle the embryo strongly in the uterus after IVF, Nurse comes to home to put the injection everyday for one month on her thigh and hip side. Due to this injection she underwent sever pain and itching in the body. Meanwhile, the physician advised her to sleep in the day time but due to the pain she cannot sleep. Also, they asked her to take the tablets till the delivery for thrice a day and after taking the tablets she took vomit every day. Also, she aware of the tablets (Livergen) and syrup but she doesn't know what was the purpose of the tablets and syrup. Monthly twice she went for the routine check up for the wellbeing of the child.

From the beginning they informed her delivery should be cesarean but actually the physician 90 percent expected normal delivery also to get normal delivery they given trips and injection due to that they had waited from morning six to evening six but it was not happen so decided to go cesarean section. She had delivered female baby on September, 2014 and she has not seen the baby and the baby was 3.5kg. After delivery they asked her to feed the baby but she don't have milk secretion so they used sucker to get milk. After handed over the child, she was admitted in the hospital for post delivery care. After delivery to still she was not able to eat normally and after eating immediately vomit will come like pregnancy. Alas, she has pain on her neck, legs and nerves and migration will occur often during walking.

Due to health complications and abuse, she was not interested to do surrogacy for second time even they are ready to offer more money because first time she don't have aware on surrogacy when she underwent. Initially they are spokes with her fruitful words but during the pregnancy she has faced more problems such as teased and abused emotionally by the relatives of intended parents and instead of the healthy foods they provide milk with more water, the food doesn't have any taste and gives the spoiled foods except in the beginning of five months.

She practiced surrogacy for the empowerment of her child but she got affected from physically and mentally and at some extent plan to die. Before pregnancy she has abused by alcoholic husband but now because of her helping tendency she has further abused in the name of money and baby. She was economically struggling before and even after underwent surrogacy. She also shared that, if her family members came to know that she is going to surrogacy they will not allowed her to undergo and because of the confidentiality she has faced problems. So she share that, she never ever refers any women to do surrogacy.

Case: 2

Ms. Nirmala, (name changed), 30 year old, she was married women from Chennai and she has studied up to 8th standard. She was a cook and her husband working as a painter. She has two children and four members in the family. Six months (May, 2015) before she delivered a boy baby to the intended parents through surrogacy.

Initially she doesn't have any idea about surrogacy and she came to know about surrogacy from print media (Daily news paper advertisement). Due to poverty, alcoholic husband, to full fill the basic needs (house rent) and to overcome economic problem of her family she decided to become surrogate.

Also, she thought will get benefits from each other through surrogacy. Very first her partner not allowed practicing surrogacy because he has wrong notions about surrogacy. Before practice surrogacy she has more doubts, those doubts was clarified by the associations and they convinced, practicing surrogacy was not a wrong thing and it's a pleasure thing. To practice surrogacy she was stayed in a rented home with her daughter in Chennai. Often her husband will come and meet her during surrogacy and she spent her time to take care her daughter, reading books and watching television.

In the contract she, her partner, association staff and hospital staff put signature. She don't know that, whether the intended parents put signature in the contract. In the contract mentioned regarding i. fixed amount ii. Monthly will give five thousand iii .processing fee paid by intended parents iv. should not ask to meet the intended parents v. should not have relationship with husband etc., at last all the clauses are they red in front of them and got sign from them. None of them forced her to practice surrogacy; she and her partner decide to do surrogacy. To undergo surrogacy they fixed Rs: 3Lakhs and monthly they paid Rs: 5k and before that she got some amount in advance at the end they paid the balance amount without any cheating. She has invested partial amount for her daughter's future as fixed deposit.

When she decided to do surrogacy at that time she has nervousness and afraid about the process. When she saw the hospital infrastructure at that time she was stunned and reluctance to speak in the reception of the hospital. From the beginning to end she does not met the intended parents. She was stayed in rented home which was arranged by intended parents. During the surrogacy process she went monthly check by car at the place the neighbors are enquired like you are facing economic problem but for check up you going by car? How can you go for check up such a big hospital? Usually people will go check up three months only but you are going monthly once? Why are you taking number of tablets? For all the above questions she replied like she has problems in her uterus due to this the doctor asked her to take more tablets for treatment and she is having more year difference between second and third child so taking special treatment. Often the house owner will come and ask more questions so she avoid to meet house owner she act like sleep and also she avoid to speak with her.

During pregnancy her partner done the entire house chore and she took complete rest. She always active before surrogacy but during surrogacy she was sluggish and taking rest was tough job to her also maintaining the confidentially of surrogacy makes her feel guilty during surrogacy.

Her mother in law came to know that from other person that she has practicing surrogacy. Due to surrogacy she faced problems from her relatives. Also they mentioned, she was characterless and doing like sex work that's why she was not contacted us.

Her daughter came to know that baby was growing in her mother stomach but after delivery she informed to her daughter that was not baby and mother affected by some tumor.

Initially they had done complete body check up then they started to do ET but the ET was failure first time and few days later again they do second time ET but this time it was success so they asked me to stay hospital for fifteen days and she felt happy for the success of ET. For ET they not used her egg, they got the egg from egg donor and in ET they transfer two embryos but only one was successfully settled. Due to her negative blood group every 24 hours they did blood test. Every month the intended parents provide nuts and fruits through the agents. Also, for festival the intended parents provide dress and snacks for her daughter. The surrogate pregnancy experience was good to her because she left her daughter during the delivery time, remaining months they are together and her relationship with husband was harmonious.

During surrogacy she has affected by pregnancy sugar because of this physician asked her to take diet. Every day of First quarter months took injection and tablets and also due to her negative blood group she asked to take some special treatment. Till the delivery she has took tablets for the well being of the child. Per day she had five to seven tablets and an injection. Due to the injection she has faced pain in her body up to six months. Also, she has experienced the movements of the child and at that time she thinks that it's not her child and it will leave her after delivery. Except partner family members are not allowed to meet her. After handed over the child, some days she think about the child later she convinced herself thinking of her own child. Now she wishes to see the face of the child even from the snaps. After delivery to stop milk secretion they give tablets to her because of this she got fever.

After the delivery she wish to see the child, want to know the growth of the child and she expect to talk with intended parents over phone. She also wish to do surrogacy second time but she expect the money up to 5 lakhs. She also shared that, the intended parents should not feel to bring the money as burden and they should give money with whole heartedly. Meanwhile the intended parents should not consider only money they also think about our health also. It's also good not allowed to see the face of the child otherwise she may refuse to leave the child and it will also affect her emotionally. The government should legalize the surrogacy and the women who was suffering economically will allow to do surrogacy and widow, divorce and separated from partner those women should practice surrogacy.

Case: 3

Ms. Tamilselvi (name changed), 26 year old married from Chennai, she has one boy child and she got married by her own choice without parents support. They are completed their family but her partner wish to see his wife with pregnant to do cultural rituals. In this scenario she came to know about surrogacy from print media (daily news paper). Also she has curiosity to know about surrogacy so she informed her partner regarding and interesting in surrogacy. But her partner refused to practice surrogacy because he felt it may affect her health but she convinced her partner and she took him to the association to get more details of surrogacy.

Based on the given information she first donated her egg later they had discussion with intended parent's assistant to practice surrogacy. Her husband met accident to safeguard her partner she got owe, to repay the loan she has decided to undergo surrogacy and help to fulfill the desire of the childless couples.

After the discussion the assistant of intended parents took her hospital, there did complete master check up later they started to ET. But the result may come 15 days later and the first ET was failure so they left home. One month later again she went to hospital to do DNC then they started second cyclic process. First time she was stayed in the hospital so the second time the doctor asked her to send and stay in home to get positive result because she may get depression in the first cyclic process so it was failed. Fifteen days later she went to hospital at that time she worshiped to get positive result, as per her desire, the second cyclic process was positive. The positive result makes her happy because first process makes her feel very worst and guilty because first failure makes the money as useless. For a women, the doctor can try twice but if the commissioning parents refuse to try second time then they women will not allow to try again, it's depends upon the intended parents. First time to get the positive result they put fifteen oil injection on her hip. Due to the power of the injection it creates tumor in the hip; those tumor was dissolved with the help of ice cubes. But the second time ET process doctor divided the fifteen injections as each five. During ET they did not disclose the number of embryo transfer. But after the positive result they declare that, they are placed three embryos and except two, one embryo was removed by the doctor. She was happy because of the failure of first time but second process gave two healthy embryos. The removal of one embryo cause internal bleeding but the bleeding does not affect other embryos. After this success, she feared about to deliver the twins healthy. Meanwhile, the doctor advised her like no need to worry, 8 months (32 to 34weeks) enough to carry the child also based on the growth of the child cesarean will be planned. Later, the doctor prescribed some tablets asked her to come one month later for further check up.

Based on the success of the ET, the contract was signed between them, in the contract she and her partner, advocate, doctor and intended parents put the signature. Before signing in the contract lots of clauses are mentioned in the contract and those calluses are red by her and her partner, also copy of contract not given to her.

Due the internal bleeding often they transfer blood to her. The entire pregnancy went smoothly and based on the growth of the child the delivered the baby in the month of eight by cesarean. The baby was under weight so they took the baby to other hospital to put in incubator. After delivery she has stayed in the hospital ten days for post delivery care later they checked the stitch and pulse of her and then they allowed her to go home.

During the pregnancy they asked her to eat nuts, fruits etc but due to nausea she refused to eat because of this she got scolded by husband, doctor and agent. Also she faced pain in the body because of the injection. In addition, due to her weakness of health they often put trips, to put trips the vein was not easily observed by staffs so it's often injured her hands. Following the delivery of twins they give her 4 lakhs (3 lakhs for one child, for twins one more lakhs) as single payment; the entire amount was used to repay the loan. Before pregnancy she has strongly kept in her mind that, the baby was not her own, should hand over after the delivery. The thought process makes her strong after handing over the child. She would like to practice surrogacy for second time to help childless couple without compensation.

Case: 4

Ms. Jaheera (name changed)*, 29 year old divorced women, she was living with parents and she having two girl child and seven members in her family living in rented house. She has completed high School.*

Before surrogacy she has dept and she doesn't have any economical work so she thought to earn from some work, to repay the loan and save her girls future. But, before surrogacy she came to know about surrogacy she had knowledge on test tube baby through the television serials and News. In this scenario she came to know about surrogacy through print media advertisement. Once red add, she immediately contacted the advertised person over phone and on the other side a lady explained about surrogacy to her. After the conversation she had discussion with her parents about the surrogacy but her parents refused to undergo surrogacy and they mentioned that, her children needs their mom, her health may affect so no need to practice and asked her to stop the process. With her interest to practice surrogacy she has convinced her parents like, she has to earn for the future of her child finally one day she went to the association with family members. In the association they explained each and every step of the surrogacy process.

Later, she donated the egg and one year later the association contacted her over phone and they informed like an Islamic couple need Islamic surrogate. With the help of the association both are met together, after that they left to hospital to start the surrogacy process. For the embryo transfer they are not used the intended mother egg, instead of that, they got egg donation from someone not from her. In hospital they took the entire necessary test and after the completion of the test one day they transfer the fertilized egg (ET). Later, they asked her to come after fifteen days for the confirmation of the embryos growth. As per the direction of Doctor she went hospital fifteen days later and after the checking process they informed her embryos have 50% growth and from the three embryos one embryo growth was good. Scan has done after three weeks and confirmed the pregnancy. After confirmation, monthly once she went for checkup up to three months and vomiting has started in the first month and *it's not stop up to four months. After three months of ET, in month of five that's 25th weeks she delivered the baby by cesarean because the heat beat of the child and the amniotic fluid also good in condition. She has delivered a female baby and she has delivered baby eight months (May, 2014) before from the date of interview. The child got her proximity and they asked her to feed the baby but she refused to feed her because she believes that mother-child bond was creates through breast feeding. The intended father celebrates the birth of the female child like his mom born him like as daughter but the mother of the baby expected boy baby. After delivery, when the relative comes to visit the baby at that time she shifted to the next door of the hospital.*

Nearby hospital she and the intended parents are stayed in the rented home till the delivery. During pregnancy she has had more number of tablets like as food, she shared that, tablets are enough instead of food, it's filled her stomach. Due to tablets and injections often she faced migration. First week of pregnancy she had continuous vomiting even she can't intake water. During pregnancy she has used to spend her time to speak with the indented mother and done prayers and her parents are not allowed to visit her so she used phone to contact them, her parents and relatives are known that she underwent surrogacy but her children and neighbors are not aware that. She informed her children that, their mom is going to serve for old age people and in between cant able to come. At pregnancy she has spoken twice a day with her children.

In the agreement she, her parents and association put signature also her brother gone through the agreement and she had doubts those doubts are clarified by the association. She demanded four lakhs but they refused to pay 4 lakhs because for single child fixed amount was three lakhs only. The amount was paid as two installments such as 2 lakhs, one month later another 1 lakh paid by them. They are not provided the copy of contract to her.

Before surrogacy she has afraid for cesarean so she has informed the doctor, her two previous deliveries are normal but they replied like normal will occur sometime. But during the time of delivery she came to know that, they will not allow normal delivery and gives hope and confident to do cesarean. Through the check up, the delivery date has fixed by the doctors. For the operation, doctor put injection on her hip to make skin as coldblooded and then they tied her eyes using block cloth. But she felt the process like tearing the skin, taking out the child without pain. After taking out the child, the baby was cried.

Her children missed their mom because of this often children health got affected also she has missed her children's proximity and some time she cried badly. When she missed her children she wished to quit but the contract stopped from her thought. Often she thinks like, she doesn't have any other option till delivery and she should be as slave. Normally, she has afraid to be alone and when she facing the thoughts at that time she used to cry alone. After delivery, her children not accepted their mom after a long gap between them; it took some days to become normal. She used half of the amount to repay owe and partial amount was she used for children's future as fixed deposit.

In pregnancy phase, due to vomiting she avoid to eat foods so the intended parents gives threat like if anything happened to our child we will not leave you but she shared with them what can I do, my body not accepting to eat. She also shared the threat with association but they replied like it's their child, they are spending more money so they will said like that only. First five months only they are not allowed her to work after that she has done all the domestic chores because in between the intended mother affected by typhoid fever. Six month later she has started to do prayer (Namaz) on cot. In her home, her mother not allowed to do domestic chore but here she has done all the chores because to make delivery as normal. Intended Parents not allowed her to eat what she wishes but she don't like to eat their preparation so she has prepared food for her own. Every day they turn on television at 08.00am and it will turn off at 03.00 am by intended mother due to this she can't able to sleep peacefully. Sometimes, she felt pain on her stomach, to get relief she used apply oil on her umbilicus at night. The baby was too active during pregnancy and often felt the mobility of the baby because of this sometime she woke in between sleep at night. Till delivery she felt like when I got relief from pregnancy and when I go to home to meet her children. One month gap her surrogacy pregnancy, her younger sister also conceived so she has eagerly wishes to saw her sister. In pregnancy she was not talk with the baby but she saw the baby and after delivery the doctor advised her to eat iron tablets for six months to improve health but the intended parents not provide tablets so the association bring one month tablets to her. To stop milk secretion the doctor advised to have tablets. She doesn't face any complication after delivery and her menstrual cycle also as usual. Very first she has believed that, the intended parents are good but they are too miser.

Case: 5

Ms.Ambiga (Name changed), 37 year old married women from Chennai. She has seven members in her family and four children (3 female and 1 male) and living in rented home. Her husband was an auto driver she was a home maker. Her husband addicted for alcoholism so the income was not sufficient to run the family. Her children completed middle school, now they are working for the family. One of her girl got married; another one got engaged and remaining male and female working in nearby Soda Company. With the help of children she was running the family. She got married at age of fifteen, she not completed schooling and she knows only to put signature.

Before surrogacy she got loan from money lender with huge interest, to repay the loan she underwent surrogacy but before surrogacy she has negative opinion on surrogacy and someone nearby her house got child from surrogacy at that time she was teased the women but she has feel guilt for that because of her unaware. In her nearest area, there was an old lady broker introduced her about surrogacy and she took her to hospital to practice surrogacy. Before surrogacy she has not donated her egg but she practiced surrogacy. During surrogacy she has convinced herself like there are number women's affected due to the childless and they are abused by their in-laws with such scenario she has shared about surrogacy with her children and she got permission from them to practice surrogacy. But her husband refused to undergo surrogacy because he had wrong notion about surrogacy also he beat her to avoid. Few days later she took her husband to the hospital and in the hospital they explained about the process of surrogacy after that her husband gave permission to undergo surrogacy. Through, throughout the process she was stayed in the hospital till the delivery and in the hospital they arranged separate room for every surrogate and provide healthy food also they are not allowed to speak with others often. Her children will come to meet her until delivery and she has spent her time watching television. Each and every room in the hospital they placed separate television for everyone. To initiate process in the hospital they doctor got sign from her and her partner also she was not aware that, who are all signed in the contract and what are all the clauses mentioned in the contract and they are not provide any copy of contract to her. For the pregnancy they had fixed 2 lakhs but she got only 1.5lakhs the remaining amount she has not received so she went to collect the balance money from the doctor but she came to know that the balance amount was collected by the broker lady from the doctor and she was cheated by the lady broker. To get the balance money she was tried to contact the broker lady but the lady was shifted her home. Also she mentioned that, if she has done surrogacy through association she may not be cheated by anyone. She does not have any problem in her family and her relationship with her husband was harmonious because there is no hide and seek. She has not aware of the intended parents and delivered baby also there is no contact as per the agreement. In some extend after the delivery if the intended parents refuse to accept the baby she has ready to take care the baby and she has no hesitation and guilty after the surrogacy because she thought she has given happiness and meaning to someone's life. In the pregnancy period she has missed her son's proximity and she has referred her friends to practice surrogacy. She wishes to the see the baby and wants to know the growth of the child and every year, her children remember the birth of the child and celebrate. Her complete surrogacy experience was good except the money cheating but after the surrogacy she has faced some health hazards such as her menstrual cycle was not stopped due to severe bleeding she went to the hospital for treatment but in the hospital ask her to wait from morning to evening unfortunately in surrogacy, they treat her in good manner.

Appendix -II

Interview Schedule

No.: **Interview Date:**

Socio – Demographic and Economic Profile

Name of the Participant:

S. No	Questions	Answer	S. No	Questions	Answer
1	Age		11	Size of family	
2	Education		12	Age at Marriage for SM	
3	Marital status		13	Age at marriage of partner	
4	Religion		14	Type of marriage	
5	Region		15	Savings	
6	Community		16	Dept	
7	Caste		17	Spouse Occupation	
8	No. of Children		18	Spouse Income	
9	Occupation		19	Annual Income	
10	Type of family		20	Monthly Expenditure	
11	Age at Surrogacy		21	Alcohol Expenditure (Monthly)	

Research Questions

Questions	*Answer*
Familiarity with the commissioning parents prior to surrogacy	1. Known Before 2. Unknown
Attitudes towards surrogacy when first heard	1. Positive 2.Neutral 3. Negative 4. No Ans.

Factors Associated with surrogacy	
Factors that influence the surrogacy Decision	**Forced Factors** 1. Livelihood Pressure 2. Poverty/ Unemployment 3. Lack of supportive earning members 4. Large size of the family with dependents 5. Low income realized in wage operations 6. Huge indebtedness 7. Family expenses- medical, educational and marriage of the daughters, 8. Others
Reasons to consider the option to become surrogate	1. To Help childless couple 2. Opportunity knocked the door 3. Attractive money 4. To realize a huge money 5. Improve the status by taking surrogacy as a profession, 6. Others
Source of information	1. Media, 2. Doctors 2. Peers Friend/Family members 3. Agents and Intermediaries 4. Awareness of surrogacy and its prospects
Method of self convincing a women to become surrogate	1. Empathy towards childless couple 2. Attractive Money offer 3. Support or pressure by family Members 4. Threat of huge debt 5. Agents convincing on huge money transaction pointing to economic vulnerability 6. Scientific information and convincing about no physical contact 7. Demo of Early Models and peer convincing
Method of external convincing a	1. Creating an Empathy towards

women to become surrogate	childless couple 2. Attractive Money offer 3.Threat of huge debt 1. Support or pressure by family Members 2. Agents convincing on huge money transaction pointing to economic vulnerability 3. Scientific information and convincing about no physical contact 4. Demo of Early Models and peer convincing
What are the coping strategies adopted to prepare to stay in surrogate hostel	1. Take the children along if they are babies 2. Friendship with other surrogate mothers 3. Regular contact with members of family 4. Periodical visit to home with the permission of the doctors 5. Involved in regular physical exercise, watching TV, reading books, hearing music, performing prayers etc. 6. Making the family members to visit regularly 7. Preparing the mind to stay aloof till the baby is born 8. Explain thoroughly to members of family and relatives and stay comfortable 9. Feeling of guilty by leaving the children and convince the children on the need, 10. 10. Others
Parties included in the written contract of surrogacy	1. Surrogate mother 2. Surrogate's Husband 3. Commissioning Parents 4. The Clinic, 5. Infertility physician, 6. Advocate, 7. Agents, 8... Psychiatric/psychologist/counsellor

Issues addressed in the written contract	1. Physiological Testing 2. Kind of prenatal care 3. Extra benefits provided by commissioning parents 4. Type of implantation/insemination; number of tries 5. What if surrogacy mother cannot get pregnant 6. Sex of the baby 7. Possibility of miscarriage 8. Extent of supervision 9. Compensation 10. Arrangements about child procreation 11. Arrangements once the child is born
Family forcing the surrogate to continue the surrogacy as a commercial activity. **If yes what background?**	1 Alcoholic Husband 2 Large families with more dependents but no adequate earning 3 Households which required finance for business of the husband 4 Families which wish to have a better economic status 5 To help childless couple who may be relatives or friends 6 To help the childless couple from abroad and attracted towards such contribution
Decision regarding abnormality found in feotus [if you have faced abortions]	1. Consider abortion 2. Doctor's advise 3. Convincing commissioning couple 4. Empathy towards surrogate by commissioning couples 5. Other Reaction 6. Not applicable
Reaction/consequences- if abnormality is found after the birth of the child	1. Commissioning parents would accept the child 2. Surrogate mothers needs to take the child 3. Clinic will find a solution; child stays there 4. Blaming the surrogate for not being alert 5. Not paying the full amount

	6. Request to bear another child 7. Other reaction 8. Mental and psychological disorders 9. Frustration and physical exhaustion. 10. NA
Financial provision for surrogate mothers if the pregnancy was terminated	1. Full payment regardless of the stage 2. Half the amount the agreed on as per the stage payment 3. Negotiating for the minimum amount the intermediaries 4. Negotiation for the maximum amount by surrogate showing the economic needs 5. Cheating with minimum payment 6. No post abortions treatment 7. Disregarding the health of the mother and no compensation 8. No money at all, 10. Other
Amount of money received by Surrogate mother	1. Below 2 L, 2. 2. Between 2.1 L to 3 L 3. Between 3.1 L to 4 L 4. Above 4 L, 5. 5. No Response
No of Instalments	1. Five 2. Four 3. Three 4. Two 5.One
Frequency of contact between the surrogate mother and the commissioning parents during the first trimester	1. Couple of times a month 2. Once in three months 3. Approximately once every month 4. As an when needed 5. No contact at all as per the agreement
Relationship of the surrogate mother with her husband at the initial stage of surrogacy	1. Harmonious 2. Dissatisfaction/coldness

	3. Major conflict or hostility 4. No discussion and interaction 5. Heartedness by Husband, 6. 6.Others [specify]
Place of stay for the surrogate mother during her pregnancy	1. Home 2. With friends/family 3. In rented home 4. In nearby village, away from home 5. Surrogate hostel 6. Hospital 7. Place chosen by the Commissioning parents 8. Any Other
Nature of contact between the surrogate mother and the clinic	1. Approached by the Clinic 2. Heard about it from acquaintances 3. Simply visited for general health check up and offered by the Doctor 4. Peers who are in the field introduced to the clinic 5. Add in the media/social media/Internet about the market for surrogate 6. Agents 7. 7.Opportunity came on its own,
Types of emotion experienced by the surrogate mother before pregnancy	1. Trust 2. Joy 3. Fear 4. . Surprise 5. Anger 6. Sadness 7. Disgust 8. .Anticipation
Decision-making for handing over of the baby – date, time etc.	1. Determined by commissioning parents

	2. Determined by clinic 3. Mutual agreement 4.....Decided by the joint discussion.
If the commissioning parents refuse to accept the baby	1. Clinic would accept the child 2. Surrogate mother should take the child 3. Clinic will find a solution till then the child stays there in the clinic 4. Given to cradle baby scheme 5. Given to childless other couples 6. Will be handed over to the intermediary 7. Others if any, 8. Not applicable
Consequences of practicing surrogacy on family life	1. Children getting distanced 2. Husband becoming friendly with surrogate mother [husband of commissioning parent] 3. Loss of contact with friends/ family members 4. Disagreements with friends/ family members 5. Household routine get affected 6. Ill-talk among neighbours 7. Separated from Husband 8. Issues with In-laws 9.Issues of Parents
Purposes for which "surrogate money" is used	1. Basic needs of the Family 2. Higher education of children 3. Medical treatment of the family members 4. Reconstruction of existing house 5. Building new house/ purchase of land 6. Savings for daughter's marriage or meeting marriage expenses of wards 7. Education 8. Repayment of loans and relief from Money lender 9. Invested in self employment/business

The worst part of being a surrogate mother	1. Problem in relationship with couple, 2. Leaving the baby(act of), 3. The secrecy involved(lying), 4. The experience of the birth: long, 5. There has been no worst part 6. Mental and psychological pressure of leaving the child 7. Cheating from making full payment 8. Ill talk by members of the family, neighbors, friends – misunderstanding surrogacy
What are the facilities you felt that you missed or sacrificed?	1. Mobility to go shopping, park etc 2. Not able to visit temples/church etc 3. Routine upset at the home front 4. Loss of Children's proximity 5. Loss of meeting sexual needs of husband 6. Not able to be normal and relaxed 7. Physical inactiveness 8. Not able to continue the hobbies 9. Feeling of being house arrested and failure to be on my own 10. Feeling of being giving up social norms for the sake of family economic vulnerability
Are you able to have your routine followed? Y/N If no, what the routine you miss?	1. Visiting parents once in a month 2. Visiting temples 3. Day today household work 4. Proximity with husband 5. Preparing special food to children on holiday 6. Going to cinema, 7. Participating in village festival 8. Missing the weekly fasting 9. Going to temple by foot 10. Celebrating the festival with all joyfulness
How did you feel during pregnancy	1. Felt like being tied up with a

till deliver and handed over the child	heavy agreement 2. Felt burdened 3. A feeling of disowning 4. No liberty to be on my own 5. Felt like how different the experience of being bearing the child for my own and for others 6. Always a pressure to adhere to standards, rules and pressure to deliver the child 7. Can't be explained in words 8. Any other
Issues in Family	1. Non-co-operation of the husband 2. Ill perception of the In-laws 3. Wrong notion about surrogacy in the society 4. Lack of understanding on the process of surrogacy by the members of family and society 5. Equating surrogacy to sex work 6. Children missing the mother for a long and the kind of behavioral change 7. Abuse by the family members post surrogate 8. No compensation for post natal surrogacy and reproductive complications
Comforts at the Surrogate Hostel	1. Excellent personal care treatment 2. Regular Nutritious Food 3. No household Responsibility 4. Complete relief from day today quarrels 5. Health assistants concern and support, love and affection 6. Sophisticated and Clean environment with all facilities 7. Immediate health care facilities and attention of the doctors 8. Provision to continue to have hobbies, temple worship, watching TV etc

	9. Concern from the Doctors, commissioning parents, peers etc
	10. Special treatment which may I never experienced in life.
Issues in the Surrogate Hostel	1. Gossiping among the inmates
	2. Finding fault with any deviations from the set norms
	3. Treatment as patient rather than a pregnant mother
	4. Verbal teasing and ridiculing by the nurse and other inmates
	5. Ill treatment by intermediaries on payment to be made
	6. Doctors not able to give freedom to mobile
Type of Psychological and emotional issues faced mentally	1 Burden that they will be bearing child for others
	2 Away from home for ten months and it pain
	3 Difficulty of convincing the family members the scientific process
	4 Exclusion or abuse by the in- laws and relatives.
	5 Wrong perception about surrogacy and equating it to sex work
	6 Disowned by in laws or parents
	7 Failure of cordial relationship with neighbors
	8 Others looking down upon while walking on the street 9. Others [Pl specify]
Mental Health Consequences	**Please tick whichever is applicable**
Frightened of all Noises	1. Severe 2. Moderate 3. Mild 4.NA
Depressed	1. Severe 2. Moderate 3. Mild 4.NA
Anxiety	1. Severe 2. Moderate 3. Mild 4.NA
Unable to Eat	1. Severe 2. Moderate 3. Mild 4.NA
Cannot Sleep	1. Severe 2. Moderate 3. Mild 4.NA
Low Self Esteem	1. Severe 2. Moderate 3. Mild 4.NA
Sexual Dysfunction	1. Severe 2. Moderate 3. Mild 4.NA
Feel like Killing Myself / Suicidal	1. Severe 2. Moderate 3. Mild 4.NA

Intensions	
I get Angry unnecessarily and Beat the children	1. Severe 2. Moderate 3. Mild 4.NA
Feel like Crying	1. Severe 2. Moderate 3. Mild 4.NA
Mental Restlessness	1. Severe 2. Moderate 3. Mild 4.NA

S. No	*Questions*	*Yes*	*No*
1	Did Anyone force you to act as a surrogate		
2	Any mediators/agents involved in this whole process		
3	Family member support it		
4	Informed to your children about you were acted as a surrogate		
5	Your children support you?		
6	You shifted your family since you are a surrogate		
7	Good relationships with your husband since you are a surrogate		
8	Good relationships with your children since you are a surrogate		
9	You are referred friends/relatives to act as a surrogate		
10	Will you refer your friends/relatives to act as a surrogate		
12	Physician asked you to take any treatment		
13	Do you feel any difficulties, during the treatment/check up		
14	They gave any rules and regulations to act as a surrogate		
15	They are asked any permission from your husband, &your family		
16	They are asked you and your partner to sign in any contracts		
17	The contract is vernacular language		
18	You or your husband gone through the agreement/contract		
19	They gave the compensation as installments		
20	They asked you stay in the surrogate hostel		
21	You feel any difficulties in the hostel		
22	They are asked you to follow any diet till delivery		
23	They are asked you to do physical work		
24	They are asked you to do yoga or any exercise		
25	They are asked to take injections or tablets		
26	Have you feel any changes in your body due to the tablets ,injections		

27	Have you sleep in the day time
28	They are asked to hear any traditional music, read books
29	You feel growth of the child
30	You feel child movements
31	Have you think that it's not my child
33	Have you speak with the child
33	Have you eaten nutritional foods (ex. Protein powder/ biscuit)
34	Physician allowed you to meet the family members
35	Physician allowed you to go your home
36	Have you feel any changes in your body after surrogacy
37	Your menstrual cycle is normal before surrogacy
38	Your menstrual cycle is normal after surrogacy
39	Have you seen the kid face after delivery
40	Have you seen the kid Face
41	Would you like to see the kid face
42	After cesarean, have you feel any changes in your body
43	The physician put any injection to stop milk secretion
44	Have you have milk secretion, after delivery
45	Have you feed the kid
46	Have you feel breast pain
47	Have you got the full currency from them
48	Any resistance faced by the surrogate mother from friends/family
49	Written contract regarding surrogacy agreement
50	Copy of the written contract with the surrogate mother
51	Awareness among surrogate mother about the clauses in the agreement
52	Sex determination tests are conducted on the surrogate mothers during pregnancy
53	Have you donated Egg
54	Have you felt a sense of fulfillment to this birth as surrogate
55	Did you get hurt when others talk ill about you
56	Was it difficult to convince your family members to adopt to surrogacy

57. Did children miss you when you were away from home
58 Did your children experience any trauma when you bear the surrogacy
59 Have you got the full satisfaction of being a surrogate
60 Have you gained from being a surrogate apart from economic gain
61 Have you or your family members thoroughly read the agreement forms and terms and conditions
62 Payment in time as agreed as installments

Appendix III

Advertisement pahmplet in Tamil

குழந்தை இல்லா தம்பதிக்கு

கருமுட்டை தானம்

செய்து உதவ தகுதியுள்ள குடும்பப் பெண்கள் தேவை

கருமுட்டை தானம் செய்வது நல்லதா ?

தானத்தில் சிறந்தது அன்னதானம் அதனினும் சிறந்தது ரத்த தானம் அதனினும் சிறந்தது கருமுட்டை தானம் எப்படியென்றால் அன்னதானம், அந்த நேரத்தில் பசியை மட்டுமே போக்கும்.ரத்த தானம் ஒரு உயிரை மட்டுமே காக்கும்.ஆனால் கருமுட்டை தானம் செய்தால் ஒரு குடும்பத்துக்கு பிள்ளை வரம் கொடுத்து அந்த குடும்பத்தையே காப்பாற்றியது போலாகும்.

கருமுட்டை தானம் என்றால் என்ன ?

குழந்தை இல்லாத தம்பதியில், மனைவிக்கு கர்ப்பப்பை நன்றாக இருக்கும் ஆனால் கருமுட்டை சரியாக உற்பத்தி ஆகாது அல்லது உற்பத்தி ஆகும் கருமுட்டை தரமானதாக இருக்காது.எனவே ஏற்கனவே கல்யாணமாகி குழந்தை பெற்ற 30 வயதுக்கு கீழ் உள்ள ஆரோக்கியமான பெண்ணின் கருமுட்டைகளை தானமாக பெறுவார்கள். பிறகு அதனை டெஸ்ட் டியூப் பேபி முறை மூலம் கருவாக உருவாக்கி, அதை குழந்தை இல்லாத மனைவியின் வயிற்றில் ஊசி வாயிலாக பதித்து, குழந்தை பெற செய்ய முடியும். இப்படி ஒரு பெண்ணின் மலடி என்ற அவப்பெயரை நீக்க உதவுவதே கருமுட்டை தானம் ஆகும்.

கருமுட்டை தானம் செய்வதால் என்ன லாபம் ?

கருமுட்டை தானம் என்பது திருமணமாகி குழந்தை பெற்ற பெண் தன்னிடம் மீதம் இருக்கும் கருமுட்டைகளை, குழந்தை இல்லாத ஒரு பெண்ணுக்கு பெருந்தன்மையுடன் கொடுத்து உதவுவதாகும்.எனவே இதில் முதல் லாபம் புண்ணியம் ஆகும் மேலும் ஒரு முறை கருமுட்டை தானம் செய்தால் ரூ 25,000/- (ரூபாய் இருபத்தைந்தாயிரம்) வரை சன்மானமாக கிடைக்கும்.அது மட்டுமல்லாமல் உடல் பரிசோதனை இலவசமாக செய்து கொள்ளலாம்.

கருமுட்டை தானம் கொடுப்பதற்கு பணம் வாங்குவது, கருமுட்டையை விற்பது போல் ஆகாதா?

கருமுட்டைதானத்திற்கு சன்மானம் என்பது கருமுட்டையை விற்பதைப் போன்றது அல்ல.ஒரு பெண் கருமுட்டை தானம் செய்வதற்காக தொடர்ந்து சில வாரங்கள் மருத்துவமனைக்கு செல்வதற்கும் , அந்தப்பெண் செலவழிக்கும் நேரத்திற்குமான ஈட்டுத் தொகையாகும்.

ஒரு முறை கருமுட்டை தானம் செய்ய எவ்வளவு நாட்களாகும்?

சாதாரணமாக தலைக்கு குளித்த (மாதவிலக்கான) நாளிலிருந்து 15 முதல் 18 நாட்களாகும். சில சமயங்களில் ரத்த பரிசோதனை ,ஸ்கேன் செய்வது,சத்து குறைபாடு இருந்தால் மாத்திரை சாப்பிட சில வாரங்கள் என அதிக பட்சம் ஒன்று முதல் ஒன்றரை மாதங்கள் ஆகலாம்.

கருமுட்டை தானம் செய்தால் ஏதாவது பக்க விளைவுகள் ஏற்படுமா?

பக்க விளைவுகள் ஏற்படுவதில்லை .இதுவரை உலகம் முழுவதும் பல லட்சம் பெண்கள் கருமுட்டை தானம் செய்திருக்கிறார்கள் பல லட்சத்தில் ஒரு சிலருக்கு ஒரு சில நாட்களுக்கு கைகால் வீக்கம் போன்ற சிறிய பக்க விளைவுகள் ஏற்பட்டிருக்கிறது. கருமுட்டை தானத்தால் பொதுவாக பக்க விளைவுகள் ஏற்படுவது இல்லை என்றே சொல்லலாம்.

குடும்பக் கட்டுப்பாடு ஆபரேஷன் செய்தவர்கள் கரு முட்டை தானம் செய்ய முடியுமா?

செய்ய முடியும். குடும்பக் கட்டுப்பாடு ஆபரேஷன் செய்திருந்தாலும் கருமுட்டைப் பையில் கருமுட்டைகள் இருப்பு இருப்பதால் அவைகளை தானம் செய்ய முடியும்.மேலும் கருமுட்டைகள் இருப்பு இருப்பதை ஒரு சில ரத்தப் பரிசோதனைகள் மற்றும் ஸ்கேன் மூலம் கண்டறிய முடியும்.

ஒரு பெண் தன்னுடைய வாழ்நாளில் எத்தனை முறை கருமுட்டை தானம் செய்யலாம்?

குழந்தை இல்லாத பெண்ணுக்கு உதவும் நோக்கத்தில் கரு முட்டை தானம் செய்வது நல்லதுதான். ஆனால் வாழ்நாளில் மூன்று அல்லது நான்கு முறைகள் டாக்டரின் ஆலோசனையுடன் செய்வதுதான் சரி. பணம் கிடைக்கிறதே என்று தொடர்ந்து கருமுட்டை தானம் செய்வது சரியல்ல.

யாருக்காக கருமுட்டை கொடுக்கிறோம் என்பது வெளியே தெரிய வருமா?

இந்திய மருத்துவ ஆராய்ச்சி கவுன்சில் வழிகாட்டி நெறிமுறைப்படி யார்,யாருக்கு கருமுட்டை கொடுகிறார்கள் என்பதோ யாரிடம் கருமுட்டை பெறுகிறார்கள் என்பதோ தெரியக்கூடாது. மேலும் குறிப்பிட்ட மருத்துவருக்கு மட்டுமே தெரிந்த ரகசியமாக இருக்கும்.இதனால் மூன்றாம் பேருக்கு தெரியாது. எனவே தைரியமாக முன்வந்து கருமுட்டை தானம் கொடுக்கலாம். மத்திய அரசாங்கத்தின் இந்திய மருத்துவ ஆராய்ச்சி கழகத்தின் (ICMR) அனுமதி பெற்ற பிரபலமான மருத்துவமனைகளில் இது செய்யப்படுவதால் அச்சம் தேவையில்லை.

கணவருக்கு அல்லது குடும்பத்தாருக்கு தெரியாமல், கருமுட்டை. தானம் செய்ய முடியுமா?

செய்யமுடியாது.கணவரோடு வாழும் பெண்கள் கணவரது ஒப்புதலுடனேயே கருமுட்டை தானம் செய்ய முடியும்.கணவரைப் பிரிந்து வாழும் பெண்களும்,கணவனை இழந்த பெண்களும், தாய் அல்லது தந்தையின் ஒப்புதலுடன் கருமுட்டை தானம் செய்ய முடியும்.

கருமுட்டை தானம் செய்யத் தேவையான தகுதிகள்

1. 30 (முப்பது) வயதுக்குள் இருக்க வேண்டும்
2. திருமணம் ஆகியிருக்க வேண்டும்
3. நோய், எதுவுமில்லாமல் ஆரோக்கியமாக இருக்க வேண்டும்
4. ஒன்றோ ரெண்டோ குழந்தைகள் இருக்க வேண்டும்
5. மாதந்தோறும் மாதவிலக்கு சரியாக இருக்க வேண்டும் .
6. பாலூட்டும் தாயாக இருக்கக் கூடாது.
7. உடல் எடை45 கிலோவில் இருந்து 65 கிலோவுக்குள் இருக்க வேண்டும்
8. புகைப்படம் ஒட்டிய ஏதேனும் அடையாள அட்டை இருக்க வேண்டும். வயதுக்கு உரிய அத்தாட்சி வேண்டும்.
   (வாக்காளர் அட்டை,ஆதார் அட்டை,ரேஷன் கார்டு,பேங்க் பாஸ் புக்,பான் கார்டு,டிரைவிங் லைசென்ஸ் போன்றவை)
9. பாஸ்போர்ட் சைஸ் போட்டோ இரண்டு வேண்டும்.

கரு முட்டை தானம் செய்ய விரும்பும் 30 வயதுக்கு கீழ் உள்ள பெண்கள் அணுக வேண்டிய முகவரி

ஸ்டார்ட் இந்தியா (START INDIA)

கரு முட்டை வங்கித் திட்டம்
46, முதல் மாடி, புதுகுளத் தெரு,
கற்கோடி அம்மன் கோயில் பின்புறம்
வள்ளுவர் கோட்டம் எதிரில்
நுங்கம்பாக்கம், சென்னை-600034

போன் : 044-3535 3888

மொபைல் : 91715 26097 மற்றும் 74011 45353

குறிப்பு : இந்த நோட்டீஸ் கிடைக்கப் பெற்றவர்கள் தங்களுக்கு உபயோகமாகாவிட்டால் தெரிந்த பெண்கள் யாருக்காவது கொடுத்து உதவும்படி கேட்டுக்கொள்கிறோம்.

Appendix IV

Surrogate Mother Facilitation-Estimation

FACILITATION FEES/CHARGES:

- The fees for surrogate facilitation services shall be Rs. 1,50,000/- (Rupees One lakh Fifty Thousand Only) to be paid to START INDIA (exclusive of services tax if any applicable)
- For surrogate facilitation services Rs 50,000 of facilitation charge to be paid as non refundable administration charges during request and another 50,000 to be paid at the time of Embryo Transfer and remaining 50,000 to be paid after successful delivery.
- Surrogate compensation charges shall be discussed with the surrogate with the help of START INDIA
- If any pregnancy failure occurred during surrogacy treatment, START INDIA will provide alternate surrogate (Up to 2 Surrogates) without charging any facilitation fees.
- All preliminary, other medical tests costs ,Insurance, Travel,Clothing,Incidental expenses shall be borne by Intended parents
- Surrogate related Expenses Rs 50,000/- to be deposited at the time of registration.

Surrogacy facilitation Estimate Cost apart from Medical expenses and Insurance

S.No	Particulars	Amount in Rs
1	Surrogate compensation(To be finalized with Surrogate)	3,00,000-4,00,000
2	Surrogate Food Allowance(10 Months X10,000)	100,000-150,000
3	Housing allowance(10X5000)	50,000
4	START INDIA Service Charge	150,000
5	Legal, Documentation and miscellaneous Charges	30,000- 50,000
	Total	6,30,000- 8,00,000

Unexpected expenses

S.No	Particulars	Amount in Rs
1	In case of IVF cycle fails, Surrogate Compensation	20,000
2	In case of miscarriage or MTP-Surrogate Compensation	30,000
2	In case of twin birth additional compensation for surrogate	1,00,000

Appendix- V

Agreement copy

भारतीय गैर न्यायिक

एक सौ रुपये Rs. 100

रु. 100 ONE HUNDRED RUPEES

सत्यमेव जयते

भारत INDIA

INDIA NON JUDICIAL

தமிழ்நாடு तमिलनाडु TAMILNADU

AX 232016

M. Premavathi
Stamp Vendor L. No: 41/2000
184. V.K. High Road, NUNGAMBAKKAM
Chennai - 600 034.

SURROGACY AGREEMENT

Agreement is made this 7th May 2015 by and between Mrs. Age 29 (hereinafter referred to as A Surrogate) and Mr.A Age 38 (herein after reffered to as surrogate's Husband) residing at Street, Chennai – 603 and Mr. (hereinafter referred to as A Genetic Father), and Mrs.S. (hereinafter referred to as A Intended Mother) residing at No 67, Chennai – 600018.

The term A Surrogate@ refers to the woman who will undergo an embryo transfer procedure utilizing an embryo(s) created with the sperm of the Genetic Father and the egg(s) of an Intended Mother, and carry said embryos to term for birth and delivery to the Genetic Father and Intended Mother. The term A Genetic Father@ refers to the sperm donor whose sperm will be used to fertilize the eggs of a designated Intended Mother to create embryos to be transferred to the Surrogate. The term A Intended Mother@ refers to the wife of the Genetic Father. The term A Child(ren)@ means each Child(ren) born to the Surrogate pursuant to the terms of this Agreement.

This Agreement is made with reference to the following facts:

1. The Genetic Father is a married-individual over the age of eighteen (18) years who is desirous of entering into the following Agreement.

2. The Genetic Father desires to have a child who is biologically related to him.

Page 1 of 10

A. RA
ADVOCATE MS /2003
No. 17/8, JAGANATHAN
STREET, MADRAS-600 0

Front Page

Gestational

THIS SURROGACY AGREEMENT is made and entered into as of ___ day of _______________, 2015, by and among Mrs/Ms.____________________________) (herein referred to as "*Surrogate*") and Mr. ___ (herein referred to as Genetic Father) and his Wife Mrs. ____________________________(herein referred to as Genetic Mother), (Genetic Father and Genetic Mother collectively referred as Intended Parents) residing at ____________________________

SURROGATE HUSBAND NOT A PARTY

And

Dr. Samit Sekhar(herein referred to as treating physician) at the Kiran Infertility Centre located at 6-2-966/4, Street Number 10, Khairatabad, Hyderabad, INDIA.

RECITALS

A. *Intended Parents*: Mr. ____________________aged about years and Mrs. ____________________, aged about years are a Married Couple living together as Husband and Wife/ ~~living as partners~~. They desire to have a child or children biologically related to them by means of Surrogacy contemplated in this Agreement. The Intended Mother warrants that she is incapable of conceiving or of carrying a pregnancy to term, or has been advised by a physician that a pregnancy would be dangerous to her health and/or the health of any child she may conceive.

must to wife.

B. *Surrogate* ____________________, *aged about*___ *Years*, a resident of the state of India; she is capable of conceiving and bearing a healthy child. ***Surrogate*** wishes to assist ***Intended Parents*** in achieving their goal of conceiving a child. ***Surrogate*** does not wish to add to her family. The sole intention of the ***Surrogate*** in entering into this Agreement is to assist ***Genetic parents*** in creating or adding to their own family.

C. ***Intended Parents*** desire to have a child by in-vitro fertilization and for the ***Surrogate*** to carry and deliver their biological child or children as a result of in-vitro fertilization.

D. The parties understand that ***Surrogate*** parenting exists in an unsettled, little established area of the law; therefore, no warranties can be made or have been made as to the outcome of judicial proceedings which could result from the conduct contemplated within this Agreement. No warranties can be made as to the ultimate obligations, liabilities or expenses of the parties should a judicial action result from actions contemplated in this Agreement.

WHILE THE PARTIES ARE ENTERING INTO THIS AGREEMENT WITH THE INTENTION OF BEING FULLY BOUND BY ITS TERMS, ALL PARTIES HAVE BEEN INFORMED BY THEIR INDEPENDENT LEGAL COUNSEL AND UNDERSTAND THAT THIS AGREEMENT IN WHOLE OR IN PART MAY BE DECLARED VOID UPON THE BASIS OF BEING AGAINST PUBLIC POLICY OR HELD UNENFORCEABLE IN WHOLE OR IN PART. ANY PORTIONS OF THIS AGREEMENT WHICH ARE HELD TO BE

UNENFORCEABLE ARE SEVERABLE. FURTHER, NO ASPECT OF THIS AGREEMENT MAY BE ENFORCED IF SUCH ENFORCEMENT VIOLATES ANY NON-WAIVABLE CIVIL OR CONSTITUTIONAL RIGHT OF ANY PARTY TO THIS AGREEMENT.

IT IS EXPRESSLY UNDERSTOOD BY ALL PARTIES TO THIS AGREEMENT THAT THIS AGREEMENT IN NO WAY REQUIRES OR PROVIDES FOR ANY PAYMENT FOR A CHILD, OR PAYMENT TO INDUCE THE SURROGATE'S RELINQUISHMENT OF PARENTAL RIGHTS, IF ANY, TO A CHILD, OR CONSENT TO AN ADOPTION.

THEREFORE, in consideration of the Recitals, which are incorporated herein by this reference, the representations, warranties, promises, covenants, and agreements of the parties contained herein, and the mutual benefits to be derived by the parties hereunder, and with the intention of being legally bound by the terms and conditions of this Agreement, the parties agree as follows:

Kiran Infertility Centre, Hyderabad, India | **Surrogacy Agreement**

Signature of: Intended Parents__________________**Doctor**

SECTION ONE – DEFINITIONS

1. **DEFINITIONS**. Certain words used in this Agreement have the following meanings:

1.1. **Child**. "Child" means the child or children conceived, carried, and delivered by the mutual cooperation of the parties

pursuant to the provisions of this Agreement.

1.2. **Embryos**. "Embryos" here refer to the embryos created through IVF using ***Genetic Father's*** sperm and ***Genetic Mother's*** eggs.

1.3. **PHYSICIAN.** Dr. Samit Sekhar- IVF & Surrogacy Program Director, KIRAN INFERTILITY CENTRE PVT.LTD solely representing KIRAN INFERTILITY CENTRE PVT.LTD - SITUATED AT 6-2-96/4 STREET NUMBER 10,KHAIRTABAD, HYDERABAD, ANDHRA PRADESH, INDIA and no one else. All queries, complaints etc in future are to be addressed and intimated to the Physician.

1.4. **In-vitro Fertilization**. "IVF" means In Vitro Fertilization of ***Genetic Mother's*** ova at the ***Kiran Infertility Centre*** by qualified physician using a method of assisted reproduction that involves surgically removing one or more eggs from ***Genetic Mother's*** ovaries after ovulation stimulation and induction and combining it with ***Genetic Father's*** sperm and, after fertilization, replacing the resulting embryo(s) in Surrogate's uterus.

1.5. **Cryopreservation**. A technique used to maintain the viability of cells or tissue by storing at very low temperatures

(freezing) -- used to preserve and store embryos or sperm for their future use in AI or IVF procedures.

1.6. **Pregnancy**. For the purposes of this agreement the first day of pregnancy shall be considered the day of the embryo transfer procedure started for that cycle or the day of the embryo transfer, should that be the case. Confirmation of pregnancy shall be considered ultra sound detection of a fetal heart beat by the attending physician.

1.7. "***Genetic Father***". Refers to the biological/genetic father of the child providing the genetic material for IVF.

1.8. "***Genetic Mother***". Refers to the biological/genetic mother of the child providing the genetic material for IVF.

1.9. "***Genetic Pa rents***". Refers collectively to ***Genetic Father*** and the ***Genetic Mother***

1.10. ***Currency:*** The currency controlling all amounts in this Agreement shall be the US Dollars.

SECTION TWO - PARTIES OBLIGATIONS

2. **PSYCHOLOGICAL EVALUATION.** ***Surrogate***, represents that prior to commencement of the IVF cycle, she has undergone a psychological evaluation conducted at ***Kiran Infertility Centre***. ***Surrogate*** expressly waives the privilege of confidentiality and permits the release of the report or other information obtained as a result of the evaluation to ***Intended Parents***.

3. **PHYSICAL EVALUATION.** ***Surrogate***, represents that prior to executing this Agreement she has undergone required physical examination/testing performed by the attending physician including testing for AIDS and sexually transmitted diseases and that she has provided information regarding her medical history that is true, correct and complete to the best of her knowledge. ***Surrogate*** expressly waives the privilege of confidentiality and permits the release of the report or other information obtained as a result of said evaluation to ***Intended Parents***. ***Intended Parents*** agree to pay for all medical and laboratory costs incurred for such physical examination/testing.

Genetic Father represents that prior to executing this Agreement he has undergone required physical examination/testing performed by the attending physician including testing for AIDS and sexually transmitted diseases and that he has provided information regarding his medical history that is true, correct and complete to the best of his knowledge. ***Genetic Father*** expressly waives the privilege of confidentiality and permits the release of the medical report or other information obtained as a result of the report to ***Surrogate*** and ***Surrogate's*** obstetrician.

Genetic Mother represents that prior to executing this Agreement, she has undergone required physical examination/testing performed by the attending physician including testing for AIDS and sexually transmitted diseases and that she has provided information regarding her medical history that is true, correct and complete to the best of her knowledge. ***Genetic Mother*** expressly waives the privilege of confidentiality and permits the release of the medical report or other information obtained as a result of the report to ***Surrogate*** and ***Surrogate's*** obstetrician.

4. **IN VITRO FERTILIZATION.** As part of her obligations under this Agreement, ***Surrogate*** shall carry a child with ***Genetic Father's*** *sperm and Eggs of* ***Genetic Mother***, she may be required to submit to an IVF protocol in order to have resulting embryos transferred into her uterus.

   4.1. **Embryo Transfer-** Procedure shall be performed by a Physician at ***Kiran Infertility Centre***.

   4.2. **In-Vitro Fertilization** shall be performed by a physician at ***Kiran Infertility Centre***. ***Surrogate*** shall attend monitoring appointments and take all the medication prescribed to her in order to prepare for the embryo transfer, including injections she may be required to be administered. ***Intended Parents*** and ***Surrogate*** agree that no more than **5 embryos** per cycle may be transferred into her uterus.

5. **MEDICAL INSTRUCTIONS/ MEDICAL CARE.** ***Surrogate***, agrees to adhere to all medical instructions given to her, including abstention from sexual intercourse or any activity that might result in semen being introduced into her body during the cycling period or as instructed by the attending physician. ***Surrogate*** further agrees to adhere to schedule and prenatal medical examinations set by the attending physician. ***Intended Parents*** agree to not engage in any hazardous sexual activity that may cause them to contract a disease that they could pass on to the ***Surrogate*** and Child through their semen and Ova, for the duration of the contract and specifically during the semen collection period and Ova Extraction.

6. **MAINTENANCE OF PREGNANCY.** Upon becoming pregnant ***Surrogate*** shall adhere to all medical instructions from her treating physician at ***Kiran Infertility Centre***, which may include, but shall not be limited to: following pre-natal examination schedules, taking recommended or prescribed medications and vitamins; and, submitting to ultrasound scans, CVS testing, alpha-foetal protein testing and other necessary or advisable testing.

   6.1. ***Surrogate*** agrees that she will carry the embryo/foetus (referred to as "child") until delivery except as set forth herein.

   6.2. ***Surrogate*** agrees to consult with a designated high-risk obstetrician during her pregnancy from time to time and upon the advisement of her obstetrician.

Kiran Infertility Centre, Hyderabad, India | **Surrogacy Agreement**

Signature of: Intended Parents________________Doctor

6.3. **Invasive testing.** ***Surrogate*** agrees to undergo an amniocentesis or other invasive testing to detect foetal genetic and congenital defects if *treating* obstetrician concurs that such testing is medically appropriate.

6.4. **Abortion/Selective Reduction.** ***Surrogate*** hereby acknowledges that she is aware of her rights as a Surrogate Mother and her constitutional right to decide whether to abort. She also acknowledges that she has been made aware of the Laws relating to Abortion in India. If Surrogate wishes to abort, ***Surrogate's*** decision to abort must comply with constitutional guidelines. If ***Surrogate*** undergoes an elective abortion in the absence of circumstances set forth below, she will be deemed in breach of this Agreement, and Physician on behalf of the breach committed by the Surrogate shall be liable to ***Intended Parents*** for all costs incurred under this Agreement, including but not limited to medical expenses and legal fees. ***Intended Parents*** recognize that some Genetic and Congenital Abnormalities may not be detected by amniocentesis or other testing, if performed. It is the intent and understanding of the parties that ***Surrogate*** shall undergo a therapeutic abortion or selective reduction under the following circumstances after prior approval from Physician and Intended Parents:

There is no cost break-up of medical fee and legal fee. What are the cost headings that are involved.

6.4.1. If her own health is threatened in the opinion of **Surrogate**'s Obstetrician/Emergency Health Care Provider and communicated to **Physician**.

6.4.2. Upon the on-going development of more than two foetuses.

6.4.3. Upon knowledge of a Genetic or Congenital abnormality including but not limited to Down syndrome, Klinefelter syndrome within 24 weeks of Pregnancy which is incompatible with life of good quality of life and after obtaining consent of the Intended Parents.

6.5. **Information Updates.** ***Kiran Infertility Centre*** agrees to notify ***Intended Parents*** of the status of pregnancy, expected delivery date and changes thereto. ***Kiran Infertility Centre*** shall promptly inform the Genetic Parents of any medical problems, conditions or concerns that may arise during the pregnancy and also agrees to provide the ***Intended Parents*** with prompt notification of the date and time of the delivery of the child, as known, Delivery shall take place, unless in cases of emergency, at ***Kiran Infertility Centre or a designated Medical Facility by Physician.*** Physician in the process agrees to provide the *Genetic Father* the following information updates on time schedules provided below:

6.5.1. Beta HCG values: 12 days after embryo transfer; repeated on day 15 post embryo transfer. **6.5.2.** Trans-Vaginal USG: Day 21 to detect a Gestational Sac: **6.5.3.** Day 31-scan to detect Heart Beat.

6.5.4. From Day 31 till pregnancy Trans-Vaginal USG every 10 to 15 days.

6.5.5. Test to rule out Down's Syndrome: Week 12- Double Test followed by Week 18 Triple Test

6.5.6. Test to detect any gross anomalies: Week 22 to 24: 4-D scan(TIFFA - Targeted Image Fetal Anomaly scan)

6.6. **Lifestyle Prohibitions.** While ***Surrogate*** executes this Agreement, and until the Agreement is terminated either by the birth of a child as contemplated herein or by termination of the Agreement as otherwise set forth in this Agreement, ***Surrogate*** shall be residing in the Premises of ***Kiran Infertility Centre*** from the date of IVF till the date of delivery or a date specified by the consulting Physician and agrees to **refrain from**:

6.6.1. Getting tattoos; participating in any dangerous ~~sports or activities as~~ advised by the attending physician.

6.6.2. Smoking, drinking alcohol and using any illegal drugs (controlled substances);

6.6.3. Using any non-prescription drugs or prescribed medication without the prior written approval of the attending physician.

6.6.4. Undergoing x-rays (except in the event of a medical emergency) or chiropractic treatment without the prior written approval of the attending physician.

Kiran Infertility Centre, Hyderabad, India | **Surrogacy Agreement**

Signature of: Intended Parents________________**Doctor**

Agreement. ***Genetic Parents*** shall take custody of and responsibility for the Child as soon as possible after the Child's birth and after the necessary releases and/or consents have been signed by ***Surrogate*** regardless of whether the Child is healthy or ill, male or female, handicapped, retarded, etc., and regardless of whether more than one child is born. ***Genetic Parents*** have no responsibility to take custody of a child unless the DNA tests confirm that they are genetically linked to the child. If the DNA tests reveal that one or both of ***Genetic Parents*** is not linked to the child, the clinic is liable to reimburse all costs and damages.

10.1. **Parenting a Handicapped Child.** Both Parties understand the responsibility to Parent a Handicapped Child and, with regards the purpose and intent of this Agreement, have concurred that the following dispositions shall prevail:

10.1.1. Handicap discovered after a therapeutic abortion is deemed impracticable or reputed illegal or discovered after Birth. In the event an abnormality is detected too late in the pregnancy and that qualified medical doctors will not conduct an abortion or that by state law it is illegal to abort or if there is no forewarning of such abnormality until birth occurred. In the event a child is born with physiological abnormalities, the ***Genetic Parents*** shall take custody of and be responsible for the child. ***Surrogate*** shall not be responsible for reimbursement of her expenses or any monies received by the family, unless it is established that her actions were responsible for the handicap of the child.

11. **GUARDIANS OF CHILD.** In the event of the death of ***Intended Parents*** prior to the release of custody of the child to them, custody of the Child born pursuant this Agreement shall be given to the guardian named by ***Intended Parents*** in their will. In this case, it is ____________________, resident of ____________________, who is of the Intended parent

__________.

12. **EXPENSES INCURRED IN THE PERFORMANCE OF THIS AGREEMENT.** ***Intended Parents*** shall pay costs and fees of medical service providers and all of ***Surrogate's*** expenses including procedures, diets, screenings, prescriptions deemed necessary by Physician from ***Kiran Infertility Centre*** that are related to the IVF procedure and/or pregnancy. Both the parties herein have agreed onto an amount of **$32,000 U.S.D. (US Dollars Thirty Two Thousand Only)** for the entire package from IVF till the Delivery of Child, Payment Plan for the same is as mentioned below.

§ **12(A).** $15,000 US as Advance.

§ **12(B).** $2,500 US at 6 Weeks of Pregnancy

§ **12(C).** $2,500 US at 12 Weeks of Pregnancy

§ **12(D).** $2,500 US at 18 Weeks of Pregnancy

§ **12(E).** $2,500 US at 24 Weeks of Pregnancy

§ **12(F).** $2,500 US at 30 Weeks of Pregnancy

§ **12(G).** $4,500 US at the time of Birth.

Services offered in the above package include:

- Psychological-Social-Physical Screening, Travel, Stay, Food, all Medications, other expenses & Compensation for the Surrogate.
- 4 IVF/ICSI procedures including Embryo Freezing. Medication and Egg retrieval process for a maximum of 4 times for the Intended Mother. If all the three Egg retrieval

their 4th IVF/ICSI procedure with multiple embryo transfers using an Indian Oocyte donor at no extra cost.

> Eric Watterson 23/6/2015 9:06 PM
> **Comment [3]:** Is a Caucasian oocyte an extra charge?

- Delivery/Caesarian Section for the surrogate. Doctor's fees for 9 months till delivery.
- Blood work, Semen Analysis; D.N.A. Test for proving paternity in India.
- Arranging the Surrogacy Agreement between all parties concerned.

> Eric Watterson 23/6/2015 8:57 PM
> **Comment [4]:** Does this include all legal fees? Are we responsible for any additional legal fees?

- Stay: Complimentary 14 days of initial stay; Post birth stay of 2 weeks at service apartment all inclusive of 3 meals, internet, laundry and nanny. Mobile-Phone for local use. Daily Clinic Visits and Airport Transfers. Foreign Exchange. Concierge to take care of local needs and site-seeing.

> Eric Watterson 23/6/2015 9:07 PM
> **Comment [5]:** If we need to go back for 2nd time (3rd and 4th time), is the service apartment cost included?

12.1. ***Legal Representation: Intended Parents*** shall pay for their legal representation and any legal fee for the establishment of their parental rights if performed outside India. ***Intended Parents*** shall pay for all court costs, filing fees and copying fees for a judicial declaration of paternity, if required to be performed outside India.

> Eric Watterson 23/6/2015 9:12 PM
> **Comment [6]:** So this is a separate fee from what we will already be paying Kiran? If so, what is the approximate cost?

12.2. ***Paternity-Maternity testing.*** All costs and fees of medical service providers for paternity-maternity testing pursuant this Agreement, if requested to be done outside India shall be borne by ***Intended Parents***. In the event child/children fail to establish Genetic Link, ***Intended Parents*** are under no obligation for the acceptance of the child and are entitled to a full refund of all payments made under this agreement.

> Eric Watterson 23/6/2015 9:11 PM
> **Comment [7]:** Again, is this a separate fee from what we will already be paying Kiran? If so, what is the approximate cost?

12.3. ***Premature delivery/ Incomplete Pregnancy.***

12.3.1. Should the ***Surrogate*** miscarry or child die before birth in or before the 6th week of pregnancy, ***Intended Parents*** will only be obligated to pay costs and fees up until the point of miscarriage or death as outlined in this ***Section (12 A-G)*** and physician shall undertake remaining IVF Attempts and subsequent multiple Embryo Transfers arising out of IVF Attempts at no additional costs.

> Eric Watterson 23/6/2015 9:15 PM
> **Comment [8]:** What exactly does "no additional cost" mean? Does this mean we do not have to pay the 15K again?

12.3.2. Should the ***Surrogate*** miscarry or child die before birth in or before the 12th week of pregnancy and after 6th week of pregnancy, ***Intended Parents*** will only be obligated to pay costs and fees up until the point of miscarriage or death as outlined in this ***Section (12 A-G)*** and physician shall undertake remaining IVF Attempts and subsequent multiple Embryo Transfers arising out of IVF Attempts at no additional costs. In this scenario only the advance payment shall be considered and payment for 6th week shall be attributed to the initial pregnancy wherein Miscarriage or death of the child occurred and the Intended Par ent shall pay this amount for New Pregnancy arising out of remaining attempts.

12.3.3. Should the ***Surrogate*** miscarry or child die before birth in or before the 24th week of pregnancy and after 12th week of Pregnancy, ***Intended Parents*** will only be obligated to pay costs and fees up until the point of miscarriage or death as outlined in this ***Section (12 A-G)*** and will not be obligated to make any further payment. In these circumstances this agreement stands dissolved.

> Eric Watterson 23/6/2015 9:47 PM
> **Comment [9]:** What does "dissolved" mean in this circumstance? Does it mean we cannot have any more tries? If the miscarriage happens on the first of 4 cycles, do we forfeit the remaining cycles?

12.3.4. Should the ***Surrogate*** miscarry or child die before birth after 24th week of pregnancy, ***Intended Parents*** will only be obligated to pay costs and fees as outlined in this ***Section (12 A-G)***. In these circumstances this agreement stands dissolved.

12.4. ***Multiple Births. Intended Parents*** shall pay an amount of **$8,000** (Eight thousand US Dollars) in excess of the above mentioned package in case of a twin pregnancy.

12.5. ***For Emergency Procedures or Surgeries not including Caesarian Section*** performed on the surrogate during the term of pregnancy or within One Month of the delivery, deemed to be as an event related to the pregnancy the costs of which shall be limited to a maximum of $1,000 shall be borne by the ***Intended Parents***. ICSI,

> Eric Watterson 23/6/2015 9:19 PM
> **Comment [10]:** Can they be more specific as to what type of procedures?

Kiran Infertility Centre, Hyderabad, India | **Surrogacy Agreement**

Signature of: Intended Parents ______________ **Doctor**

(8) 12.6 What is [illegible] compensation of SM?
Who is responsible for payment of compensation to SM?
Separate agreement with AMMA?

limited to a maximum of $1,000 shall be borne by the *Intended Parents*. **ICSI, Blastocyst transfer and Assisted Hatching- if required is done at no additional cost. PIGD – *Pre Implantation Genetic Diagnosis* shall be done at an additional $5000; Cord Blood and Cord tissue banking shall be done at an additional cost of 1200us$** .Sperm wash to remove STD's shall be done at an additional $3000.

12.6. By entering into the surrogacy program you agree that you are aware that the clinic is responsible for the Medical care of the surrogate mother but not for her financial remuneration. You understand that you are paying the surrogate mother reasonable expenses. You consent to the clinic releasing the surrogate mother reasonable expenses to AMMA women welfare society represented by their vice president Mr. Sesha Sai, who recruits the surrogate mothers for the clinic, who in turn will release it to the surrogate mother from time to time as per her demand. You understand that the clinic is only facilitating payments to your surrogate mother as per the amount and terms and conditions decided by yourself and the surrogate mother through Amma women welfare society and the clinic will not be paying the surrogate mother any financial remuneration directly as per the ICMR guidelines.

Eric Watterson 23/9/2015 9:23 PM
Comment [11]: ? We will not have contact with the surrogate, correct?

SECTION THREE - CONTRACTUAL CONSIDERATIONS

13. **REPRESENTATIONS AND WARRANTIES.** The parties are making the following representations and warranties:

13.1. ***Surrogate***, represents and warrants to the ***Intended Parents*** the following:

13.1.1. ***Surrogate*** enters into this Agreement with the clear intention to assist the ***Intended Parents*** in their parenting the child conceived and born pursuant to this Agreement, and she has no intention or interest in parenting the child or assuming the responsibilities inherent in parenting.

13.1.2. To the best of ***Surrogate***'s knowledge and belief, she is capable of conceiving and bearing a child;

13.1.3. ***Surrogate*** has had a normal menstrual period within the last month and she is not currently pregnant.

13.1.4. All personal information provided to ***Intended Parents***, has been complete, and true to the best of her knowledge, including, but not limited to, medical information and information about previous arrest records or child abuse charges.

13.1.5. ***Surrogate*** obligations set forth in Section 2 of this Agreement are in the best interests of any child conceived as a result of the actions contemplated in this Agreement.

13.1.6. In the event of a dispute over custody of the child, the best interests of the child shall control, ***Intended Parents*** shall have standing to seek sole custody of the child.

Eric Watterson 23/9/2015 9:26 PM
Comment [12]: Why would there ever be a dispute over the custody of the child?

13.2. ***Intended Parents*** represents and warrants ***Surrogate*** the following:

13.2.1. ***Intended Parents*** enter into this Agreement with the clear intention to parent the Child conceived and born pursuant to this Agreement, and assume all obligations and responsibility for the Child regardless of whether the Child is healthy or ill, male or female, handicapped, retarded, etc., and regardless of whether more than one child is born:

Kiran Infertility Centre, Hyderabad, India | **Surrogacy Agreement**

Signature of: Intended Parents____________________ **Doctor**

13.2.2. ***Intended Parents*** have undergone all medical and laboratory tests to establish that they are free of sexually transmittable disease that could be pass on ***Surrogate*** or future Child through the AI or IVF procedure;

13.2.3. ***Intended Parents*** are capable, to the best of their knowledge to produce semen, sperm and Ova of sufficient quality and quantity to conceive a Child through the AI or IVF procedures contemplated in this Agreement;

13.2.4. ***Intended Parents*** warrant that they will not receive any blood transfusions, use contaminated needles nor engage in any activity, sexual or other, that could lead them to contract any disease that they could pass on to ***Surrogate*** or future Child through IVF procedure; should they receive an emergency blood transfusion, IVF procedures shall be suspended until their disease free status can be established;

13.2.5. ***Intended Parents*** have the resources and means to carry on their obligations as set forth in section 2 and to provide for the conception and support of the Child conceived as a result of the actions contemplated in this Agreement;

13.3. All parties expressly understand and agree that the professionals whose services have been retained under said Agreement do not in any way warrant that:

13.3.1. ***Surrogate*** will in fact conceive a child through in IVF procedures;

13.3.2. The child, if conceived, will be a physically and mentally healthy child, free of congenital or genetic defects;

13.3.3. ***Surrogate*** will comply with the terms and provisions of said Agreement;

13.3.4. ***Intended Parents*** will comply with the terms and provisions of said Agreement; or that

13.3.5. Any child born as a result of this Agreement will be deemed the legal child of ***Intended Parents***, and not the legal child of ***Surrogate***.

13.4. Should <u>Death of the Child</u> occur after the 24th week of pregnancy but before the completion of the 28th week, the ***Intended Parents*** will pay the aforementioned expenses in ***Section 12*** of this Agreement. How much?

Eric Walterson 23/6/2015 9:33 PM
Comment [13]: Then, would we still be eligible for our other tries [as long as this is not try #4]?

13.5. **No Untrue or Material Misstatements.** The representations and warranties made by ***Surrogate*** in this Agreement do not contain any untrue statement of a material fact or omit to state a fact necessary in order to make the statements herein, in light of the circumstances under which she were made, not misleading.

14. **ASSUMPTION OF RISK.**

The parties acknowledge the following:

14.1. The parties specifically assume the risk that the ***Surrogate*** procedure has been tested and has certain limitations within the State of Republic of India and any legal action initiated as a result of this Agreement by any party may constitute a case of first impression within the State of Republic of India.

Eric Walterson 24/6/2015 3:21 PM
Comment [14]: What does this word mean?

14.2. In the event a child is born with physiological abnormalities, the ***Intended Parents*** shall take custody of and be responsible for the child.

Signature of: Intended Parents________________**Doctor**

14.3. ***Surrogate*** agrees to assume all risk including the risk of death which may be incidental but not limited to conception, pregnancy, childbirth and/or postpartum complications. Said risks have been explained and are understood by ***Surrogate***. ***Surrogate*** agrees to assume all of said risks, and release Genetic Parents, including any medical professionals and others involved in any aspect of this Agreement from any legal liability.

15. ADVICE OF COUNSEL.

15.1. ***Intended Parents*** acknowledge that they have retained independent counsel to advise them on all applicable law, rights, duties, liabilities, terms and conditions as it relates to all matters relevant to this Agreement, and that neither the ***Surrogate***, her representatives nor agents have made any representations as to the legal consequences of this Agreement for ***Intended Parents***.

15.2. ***Surrogate*** acknowledges that she has forfeited her right to an independent counsel to advise her on all applicable law, rights, duties, liabilities, terms and conditions as it relates to all matters relevant to this Agreement, and that neither the ***Intended Parents*** nor their representatives nor agents, have made any representations as to the legal consequences of this Agreement for ***Surrogate***.

15.3. In the event of litigation between the parties and/or their families or estates, and/or any state, nation, municipality, governing body or party, counsel retained in connection with the negotiation, preparation and execution of this Agreement, shall not be required to represent any party. All parties will assume full responsibility for the payment of all attorney fees, and associated costs and expenses of litigation.

16. CONFIDENTIALITY.

16.1. The parties shall maintain the confidentiality of their relationship and protect the identity of all other involved parties, unless otherwise agreed upon in writing by all parties. None of the parties shall provide, or allow to be provided, any information to the media, public or any other individual regarding his/her involvement in or the terms of this Agreement. Notwithstanding, this confidentiality provision shall not prohibit any party from disclosing and discussing generally his/her own involvement in a ***Surrogate*** program with significant others, relatives or friends, but the identity of the other parties and the specific terms of this Agreement shall not be disclosed to parties other than significant others, relatives or friends unless otherwise agreed upon by all parties in writing.

17. TERMINATION OF AGREEMENT.

17.1. **Escape clause.** In the event that pregnancy is not achieved within a reasonable time considering ***Surrogate's*** age and physical condition, and after advice from the Consulting Physician from ***Kiran Infertility Centre***, this Agreement may be terminated by written notice from any party. After such termination, no further liability will result on the part of any party. However, it is expressly understood that the agreement shall not be terminated in the event the ***Surrogate*** has passed the 20th week of Pregnancy.

17.2. **Material breach.** In the event that any party materially violates any of the provisions stated in this Agreement without sufficient legal excuse, such violation shall constitute a material breach of the contract. This Agreement shall terminate forthwith at the option, and upon written notice from, the aggrieved party or parties. In the event ***Surrogate*** breaches this Agreement, Physician shall reimburse ***Intended Parents*** for all sums expended pursuant to this Agreement plus interest at the maximum allowable rate at the time the breach was discovered. In the event ***Intended Parents*** breach this Agreement, they shall be responsible for all expenses incurred by ***Surrogate*** pursuant to Clause 12 and all related sub Clauses in Clause 12.

Kiran Infertility Centre, Hyderabad, India | **Surrogacy Agreement**

Signature of: Intended Parents ____________ **Doctor**

17.3. **Remedy for breach.** Continued performance after a material breach shall not constitute a waiver, unless agreed upon in writing by all parties, and all rights and retained by the aggrieved party will remain in force. The violating party shall be informed in writing of the alleged violation and shall have a reasonable opportunity to cure the breach, if possible.

17.4. **Survival.** The representations and warranties set forth in Paragraph 13 of this Agreement shall survive termination of this Agreement.

18. **MISCELLANEOUS.**

18.1. **Entire Agreement.** This Agreement constitutes the entire Agreement between the parties hereto with respect to the subject matter hereof, and supersedes all prior and contemporaneous agreements, representations, negotiations and understandings, verbal and written, with respect to the matters covered hereby. The attached Medical Information Release and any fully executed Addendum are made a part of this Agreement by reference.

18.2. **Binding Effect.** The terms and conditions of this Agreement shall extend to, inure to the benefit of and be binding

upon the parties hereto and their respective permitted successors and assigns.

18.3. **Amendment.** This Agreement may be amended, modified or supplemented, and a right hereunder may be waived only by written instrument duly executed by all the parties.

18.4. **Severability.** If any provision of this Agreement is declared by a court of competent jurisdiction to be invalid or unenforceable, or if any provision hereof is or becomes impracticable, the remaining provisions and the Agreement as a whole shall nevertheless continue in full force and effect without being impaired or invalidated in any way, and the parties shall replace the invalid, unenforceable or impracticable provision with a valid, enforceable or practical provision which shall meet the economic aims of the invalid, unenforceable or impracticable provision as closely as possible.

18.5. **Other Documents.** The parties hereto agree to execute such other and further documents as may be necessary to

carry out the purposes hereof.

18.6. **Cumulative Rights.** The rights of the parties hereto are cumulative, and no exercise or enforcement by a party of any right or remedy hereunder shall preclude the exercise or enforcement by that party of any other right or remedy herein contained, or to which it is entitled by law.

18.7. **Arbitration and Mediation.** Any controversy arising under or relating to the actions contemplated in this Agreement shall be submitted to first to mediation.

18.8. **Jurisdiction.** This Surrogate Contract has been drafted in Republic of India, the Parties have availed themselves of professionals in Republic of India, and as such, it is the express intent of the Parties and material term of this Agreement that in the event of a breach of contract or dispute NOT related to the birth of a child(ren) or the finalization of parental rights in the Republic of India, this Agreement shall be governed by and construed and enforced with, the laws of Republic of India. The Parties further stipulate and agree that for the above-stated purposes of a breach of contract or contract dispute, that ***Indian Law*** shall govern any conflicts, interpretation and/or applications of this Agreement. For said above-stated purposes, the Parties agree that subject matter and personal jurisdiction shall be vested in the courts of

Signature of: Intended Parents________________Doctor

www.ingramcontent.com/pod-product-compliance
Lightning Source LLC
LaVergne TN
LVHW010718210926
848485LV00008B/1444/J